# Rapid Screening of Pathogens in Water

Awwa
Research
Foundation

## About the Awwa Research Foundation

The Awwa Research Foundation (AwwaRF) is a member-supported, international, nonprofit organization that sponsors research to enable water utilities, public health agencies, and other professionals to provide safe and affordable drinking water to consumers.

The Foundation's mission is to advance the science of water to improve the quality of life. To achieve this mission, the Foundation sponsors studies on all aspects of drinking water, including supply and resources, treatment, monitoring and analysis, distribution, management, and health effects. Funding for research is provided primarily by subscription payments from approximately 1,000 utilities, consulting firms, and manufacturers in North America and abroad. Additional funding comes from collaborative partnerships with other national and international organizations, allowing for resources to be leveraged, expertise to be shared, and broad-based knowledge to be developed and disseminated. Government funding serves as a third source of research dollars.

From its headquarters in Denver, Colorado, the Foundation's staff directs and supports the efforts of more than 800 volunteers who serve on the board of trustees and various committees. These volunteers represent many facets of the water industry, and contribute their expertise to select and monitor research studies that benefit the entire drinking water community.

The results of research are disseminated through a number of channels, including reports, the Web site, conferences, and periodicals.

For subscribers, the Foundation serves as a cooperative program in which water suppliers unite to pool their resources. By applying Foundation research findings, these water suppliers can save substantial costs and stay on the leading edge of drinking water science and technology. Since its inception, AwwaRF has supplied the water community with more than $300 million in applied research.

More information about the Foundation and how to become a subscriber is available on the Web at **www.awwarf.org.**

# Rapid Screening of Pathogens in Water

Prepared by:
**Hélène Baribeau**
Carollo Engineers, P.C.
10540 Talbert Avenue, Suite 200 East, Fountain Valley, California 92708
and

**Paul A. Rochelle** and **Ricardo De Leon**
Metropolitan Water District of Southern California
Water Quality Laboratory, 700 Moreno Avenue, La Verne, California 91750

Sponsored by:
**Awwa Research Foundation**
6666 West Quincy Avenue, Denver, CO 80235-3098

Published by:

Awwa Research Foundation

American Water Works Association

IWA Publishing

**DISCLAIMER**

This study was funded by the Awwa Research Foundation (AwwaRF). AwwaRF assumes no responsibility for the content of the research study reported in this publication or for the opinions or statements of fact expressed in the report. The mention of trade names for commercial products does not represent or imply the approval or endorsement of AwwaRF. This report is presented solely for informational purposes.

# CONTENTS

# TABLES

# FIGURES

# FOREWORD

The Awwa Research Foundation is a nonprofit corporation that is dedicated to the implementation of a research effort to help utilities respond to regulatory requirements and traditional high-priority concerns of the industry. The research agenda is developed through a process of consultation with subscribers and drinking water professionals. Under the umbrella of a Strategic Research Plan, the Research Advisory Council prioritizes the suggested projects based upon current and future needs, applicability, and past work; the recommendations are forwarded to the Board of Trustees for initial selection. The foundation also sponsors research projects through the unsolicited proposal process; the Collaborative Research, Research Applications, and Tailored Collaboration programs; and various joint research efforts with organizations such as the U.S. Environmental Protection Agency, the U.S. Bureau of Reclamation, and the Association of California Water Agencies.

This publication is a result of one of these sponsored studies, and it is hoped that its findings will be applied in communities throughout the world. The following report serves not only as a means of communicating the results of the water industry's centralized research program but also as a tool to enlist the further support of the nonmember utilities and individuals.

Projects are managed closely from their inception to the final report by the foundation's staff and large cadre of volunteers who willingly contribute their time and expertise. The foundation serves a planning and management function and awards contracts to other institutions such as water utilities, universities, and engineering firms. The funding for this research effort comes primarily from the Subscription Program, through which water utilities subscribe to the research program and make an annual payment proportionate to the volume of water they deliver and consultants and manufacturers subscribe based on their annual billings. The program offers a cost-effective and fair method for funding research in the public interest.

A broad spectrum of water supply issues is addressed by the foundation's research agenda: resources, treatment and operations, distribution and storage, water quality and analysis, toxicology, economics, and management. The ultimate purpose of the coordinated effort is to assist water suppliers to provide the highest possible quality of water economically and reliably. The true benefits are realized when the results are implemented at the utility level. The foundation's trustees are pleased to offer this publication as a contribution toward that end.

Walter J. Bishop
Chair, Board of Trustees
Awwa Research Foundation

Robert C. Renner, P.E.
Executive Director
Awwa Research Foundation

## ACKNOWLEDGMENTS

The authors wish to thank the following current and former staff of the Water Quality Laboratory at the Metropolitan Water District of Southern California for their cooperation and participation in this project: Anagha M. Chitre, Donna M. Ferguson, Patrick A. Hacker, Troy J. Handojo, Katrin G. Hanley, Anne M. Johnson, Jason La, Eileen Maura Jutras, and Howard Wen.

The following individuals also contributed to this study by providing microorganisms: Marilyn M. Marshall (Sterling Parasitology Laboratories, University of Arizona, Tucson), Cynthia Chappell (University of Texas, Houston, Texas), Steve Wright (Moredun Research Institute, Peniculk, Scotland), and Anne C. Camper (Montana State University, Bozeman, Montana). The research team would also like to acknowledge Jens J. Hyldig-Nielsen and Henrik Stender (formerly of Boston Probes, Inc. but now with Applied Biosystems) for the PNA probes and the technical assistance provided while investigating this aspect of the study. The advice, help, patience and support from the AwwaRF Project Managers – Emerson Lomaquahu and Ryan Ulrich, and Project Advisory Committee (PAC) – Becky Hoffman (Wisconsin State Laboratory of Hygiene), John Lisle (USGS Center for Coastal and Watershed Studies), and Cheryl Norton (American Water Works Service Company) are also appreciated.

# EXECUTIVE SUMMARY

Microbial contamination of surface and ground waters in the United States continues to occur from point and non-point sources. Population growth, land development, high volume animal farming practices, increased pressure for recreational use of lakes and rivers, and the necessity of tapping new sources of drinking water are all likely to contribute to future episodes of contamination. Such contamination of water presents a serious public health concern because of the possibility of microbial disease transmission via water. Waterborne disease outbreaks and detection of fecally-associated bacteria and enteric pathogens provide evidence that water is vulnerable to fecal contamination and can serve as a route of transmission of diseases.

Protozoan parasites such as *Cryptosporidium parvum* and *Giardia duodenalis* have been recognized as important etiologic agents of waterborne disease outbreaks. *C. parvum* is of major concern because it exhibits high resistance to chemical disinfectants at the doses applied in treatment plants, has low infectious dose, and prophylaxis or therapeutic drugs are currently not commercially available. *G. duodenalis* has historically been the most frequently identified cause of waterborne disease outbreaks in the United States. Enteric viruses such as Hepatitis A virus, Norwalk virus, and Rotaviruses, are responsible for approximately 10% of the reported waterborne disease outbreaks where an etiologic agent was identified. It is likely that many of the outbreaks for which no etiologic agent was identified (approximately 50% of all reported outbreaks) were caused by viruses but because of limited routine monitoring for viruses in water and inadequate methods the causative agent was not identified. Some *Escherichia coli* strains are capable of causing human illness, including diarrhea, by several different mechanisms. These enterovirulent *E. coli* have been implicated in several major waterborne outbreaks including a well publicized *E. coli* O157:H7 outbreak in Walkerton, Canada, in 2000. Waterborne disease data from the United Kingdom, Scandinavia, Canada, Japan, Israel and elsewhere suggests similar etiological patterns and highlight the importance of continued improvements in microbial monitoring for the water industry. Moreover, recent waterborne outbreaks have clearly demonstrated that consumption of contaminated drinking water is not solely a problem in the developing world; all of the industrialized nations of the world, with highly efficient and modernized treatment plants, continue to experience waterborne disease outbreaks.

A common factor with these microorganisms is the lack of simple and rapid methods for their detection and identification in water. Currently approved or recommended methods for detection of microorganisms in water are based on lengthy conventional culturing or microscopic enumeration techniques. These methods are not designed for high throughput or speed of analysis and the inclusion of a direct link to automation and robotic technology has seldom been a consideration. Microbial detection assays that incorporate efficient detection, the ability to process large sample numbers, and simplicity of operation are needed to serve the long term needs of the water industry. Future methods need to incorporate specific detection of target microorganisms at the species or strain level, minimal or no risk of false identification, and sensitive detection of the low levels of pathogens typically encountered in source and treated water. Ideally, these methods would also provide information in a timely manner to help prevent or minimize consumer exposure to microbial contaminants. Molecular techniques provide new avenues for detection and identification of specific microorganisms and an opportunity to address several of the problems identified with conventional methods. Molecular assays, and in particular the polymerase chain reaction (PCR), have been proposed and used for many years as alternative methods for the detection of pathogens in water. However, the standard detection end-point for these assays of gel electrophoresis followed by membrane hybridization can be time-consuming, labor-intensive, and are not readily automated.

This project developed a simple and rapid screening assay for selected microbial pathogens by adapting molecular methods for use in a convenient 96-well microplate format. High sensitivity and specificity of detection are achieved by PCR-based nucleic acid amplification with confirmation by internal probe hybridization to enhance the specificity of the assay and prevent detection of false positives from sub-optimal target amplification. Internal probe hybridization in a microplate format significantly decreases the analysis time when compared to membrane hybridization or the conventional techniques used for pathogen detection. In addition, a microplate format offers high throughput, rapid reading and potential for automation at low cost with the use of multi-well microplates and a microplate reader. Since the vast majority of water samples are expected to be negative for pathogens, a rapid screen assay allows the option of eliminating negative samples from further analysis, and conducting more lengthy and costly conventional techniques such as culturing assays for pathogen enumeration or infectivity determinations on positive samples only.

## OBJECTIVES OF THE STUDY

The overall objective of this project was to develop a sensitive, specific, robust, and rapid screening assay for the detection of pathogens in water. The project aimed to develop a simple microplate-based assay for rapid screening of selected pathogens in water: *C. parvum*, *G. duodenalis*, echovirus 1, and *E. coli* O157:H7. The specific objectives were:

1. Optimize a prototype assay comprising PCR amplification and microplate hybridization using *C. parvum* as a model organism.
2. Evaluate the adaptability of the assay to detect other organisms using echovirus 1, *G. duodenalis*, and *E. coli* O157:H7 as representatives of each type of organism (viruses, protozoa, and bacteria).
3. Further develop the assay by comparing the prototype detection assays for *C. parvum* and enteroviruses with:
    a. Other microplate hybridization approaches.
    b. Fluorescence-based detection.
    c. Peptide nucleic acid (PNA) probes instead of DNA probes.
4. Evaluate the assay using environmental water samples seeded with *C. parvum* and echovirus 1.

## RESEARCH APPROACH

In a molecular-based assay, the amplification target, the hybridization conditions, and the detection signal or moiety are the three key components that need to be evaluated and optimized. Thus, different combinations of amplification target, microplate hybridization and end-point detection approaches were systematically examined with regards to sensitivity, specificity and robustness. PCR amplification of the DNA extracted from the target microorganisms was conducted, followed by hybridization and signal detection in a 96-well streptavidin-coated microplate. Different non-radioactive haptens were incorporated during the amplification reaction either to attach the amplicons to the microplate wall or detect the amplicon-probe hybrids, depending upon the approach used. The prototype detection assay involved probe-coated microplates (PCM) and consisted of the following sequential components:

1. Digoxigenin (DIG)-labeling of amplicons during PCR;
2. Amplicon capture by biotin-labeled oligonucleotide probes coated on 96-well microplates;
3. Colorimetric detection of amplicon-probe hybrids with an anti-DIG antibody labeled with alkaline phosphatase and a colorimetric substrate.

The assay end-point (a purple colored precipitate in positive microplate wells) was detected visually for qualitative results or by using a microplate reader for relative quantification. A second assay approach involving amplicon-coated microplates (ACM) consisted of the following sequential components:

1. Incorporation of a biotin label in the amplicons during PCR amplification;
2. Binding the biotin-labeled amplicons to a streptavidin coated 96-well microplate;
3. Hybridization of a DIG-labeled oligonucleotide probe to the bound amplicons;
4. Colorimetric detection of amplicon-probe hybrids with an anti-DIG antibody labeled with alkaline phosphatase and a colorimetric substrate.

Two example microorganisms, *C. parvum* and echovirus 1, were used to determine the optimum assay conditions. Selection of these microorganisms for assay development was based on their recognized public health threat and their very different biological properties; *C. parvum* is a DNA-containing organism of concern to the drinking water industry, whereas echovirus 1 has an RNA genome and is a representative of the important enterovirus group of pathogens. Following optimization, the assay was modified for *G. duodenalis* and *E. coli* O157:H7 to assess its adaptability to a broader range of pathogens. All of these organisms have been implicated in waterborne disease outbreaks.

## CONCLUSIONS

An adaptable rapid screening assay was developed for the detection of pathogens in water. The assay was based on an industry-standard 96-well microplate platform that allowed qualitative results to be recorded by simple visual observation and relative quantification by an automated microplate reader. The assay was based on a flexible design that allowed the use of various probe formats and mechanisms of probe attachment, and the assay endpoint was adaptable to both colorimetric and fluorescence-based detection approaches. Proof of principle and demonstration of feasibility was provided by a series of experiments with four selected pathogens (*C. parvum*, *G. duodenalis*, echovirus 1, and *E. coli* O157:H7) but the assay is applicable to any pathogens for which nucleic acid sequence information is available, allowing design of specific primers and probes. Specific conclusions of this project include the following:

1. The rapid screening assay is a useful screening tool that may be used to eliminate negative samples, which constitute the majority of environmental water samples. More extensive analyses such as microbial enumeration and viability infectivity assays can then be focused solely on the positive samples.

2. The microplate format allows many samples to be processed simultaneously without increasing analysis time. Consequently, processing time and costs are substantially reduced compared to conventional culture and microscopy-based microbial detection analyses.
3. Two assay formats were evaluated involving either probe-coated microplates (PCM) or amplicon-coated microplates (ACM). Both provided similar results in terms of assay sensitivity and duration. However, the ACM format was selected as the most appropriate because it was less expensive, more readily adapted to a wide range of organisms, less sensitive to variable assay conditions, and had greater flexibility in analysis options compared to the PCM approach.
4. As the first component of the assay, label incorporation into amplicons was achieved without modifying reported amplification conditions. A single round of amplification (35 – 40 cycles) provided sufficient sensitivity and amplicon purification was not necessary.
5. The ACM approach was only slightly affected by changes in the different assay parameters. The critical steps were amplicon denaturation and pre-coating techniques, sample volume, sodium concentration during hybridization, and sufficient contact with the colorimetric substrate.
6. For *C. parvum*, a published primer/probe combination was used for the assay and a detection sensitivity of less than 10 oocysts was achieved (based on serial dilutions of hemocytometer-enumerated oocyst stock preparations).
7. With echovirus 1, the assay sensitivity was less than 0.02 plaque forming units (PFU; equivalent to less than 20 virus particles) using a primer/probe combination specific for all human enteroviruses including poliovirus, coxsackie A and B viruses, and echoviruses.
8. The assay was easily adapted for the detection of other microorganisms such as *G. duodenalis* and *E. coli* O157:H7. Published primer/probe combinations and amplification conditions were used without further modification. The sensitivity was 9 cysts and 10 colony forming units (CFU) for *G. duodenalis* and *E. coli* O157:H7, respectively.
9. Evaluations of assay sensitivity with seeded environmental water samples demonstrated sensitivities of 2 – 9 *C. parvum* oocysts (based on immunofluorescent enumeration) and 0.004 – 0.073 PFU of echovirus 1 (equivalent to 4 – 73 virions).
10. Evaluation of a commercially available hybridization-capture kit (OligoDetect) customized for *C. parvum* and incorporating the same primer/probe combination used for the rapid screening assay demonstrated sensitive and specific detection. The protocol for the commercial kit was more streamlined and so required less analysis time (6.4 h) compared to the assay developed for the project (10.8 h), although considerable optimization is still required for successful application of the kit. However, commercialization of a similar technology demonstrates the value and utility of the type of assay developed for this project.

## RECOMMENDATIONS AND FUTURE RESEARCH

The demonstration of feasibility of a microplate-based rapid screening assay provided by this project indicates that the water industry should continue to pursue and fund exploratory research into the development of innovative pathogen detection methods. Such techniques may be used for routine screening of drinking water, during outbreak situations, or to rule out (or in) intentional contamination. All of these situations require pathogen detection tools that can generate results, even if only presumptive results, in as short a turn around time as possible. Although considerable assay optimization and evaluation were performed during the course of this project, several issues that warrant further evaluation were identified:

1. The microplate-based assay was developed as a presence/absence screening tool for the detection of pathogens in water. Relative quantification was achieved through the use of an automated plate reader but an assessment of possible interferences is necessary. More extensive evaluations with precisely enumerated seed organisms need to be performed to improve the quantitative aspects of the assay and to more fully assess the sensitivity. In the case of *C. parvum*, oocysts enumerated by flow cytometry would be appropriate as seed organisms.
2. The assay was developed as a detection tool to use in combination with PCR (or other nucleic acid amplification methods) following recovery of organisms from water and extraction of nucleic acids. However, the utility of the assay would be increased by coupling it to direct nucleic extraction techniques so that potentially all pathogens can be screened from a single nucleic acid extraction thereby uncoupling the assay from time-consuming and often inefficient conventional microbial recovery methods.
3. Assay standardization, validation and round-robin testing are required to assess the transferability of the technology. In addition, the assay would benefit from streamlining and simplification in some of the processing steps, resulting in greater transferability. Specifically, further experiments involving shorter incubation periods are necessary to determine the minimum possible assay duration.
4. Seeded studies using a variety of source water types should be conducted to assess the robustness of the assay and to examine possible interferences.
5. Since the microplate assay is a detection tool, it can be used for applications other than pathogen detection. As an example, it could be used as an alternative to gel electrophoresis and membrane hybridization in cell culture-based viral and protozoal infectivity assays that use PCR or RT-PCR to detect infections. This and other potential applications of the assay should be explored in more detail.

# CHAPTER 1
# INTRODUCTION

*This Introduction/Literature Review was prepared in 2002.

Surface and ground waters in the United States continue to be subjected to fecal contamination from a variety of sources, including sewage treatment plant effluent, on-site septic waste treatment discharges, land run-off from urban, agricultural and natural areas, and leaching from sanitary landfills. With continued population growth and land development, such fecal contamination is likely to increase unless more stringent pollution control measures are implemented. Fecal contamination of water is a public health concern because of the possibility of microbial disease transmission via water used for many beneficial purposes in which there is human ingestion of or contact with the water. These uses include potable water supplies, primary contact recreation, irrigation of crops that are eaten raw, and irrigation of parks and golf courses.

Evidence for fecal contamination of surface water is provided by the detection of many pathogens in water and the continued occurrence of outbreaks of waterborne disease. For example, between 1971 and 1985, 502 drinking waterborne outbreaks of disease involving 111,228 cases of illness were reported in the United States, of which 49% were associated with groundwater sources and 51% were associated with surface water sources (Craun, 1988). During the ten-year period ending in 1990, 291 waterborne outbreaks were reported in community (43%) and non-community (33%) systems, and from ingestion of contaminated recreational (14%) and individual (10%) water sources (Craun, 1992). Untreated or inadequately disinfected groundwater was responsible for 43% of all reported outbreaks, and contaminated, untreated surface water or inadequately treated surface water was responsible for 24% of all reported outbreaks. From 1991 to 1998 more than 443,000 cases of illness occurred during 230 reported waterborne outbreaks in the U.S. (Craun et al., 2003). Of the 126 drinking water outbreaks, 87% were associated with public water systems. These outbreaks were caused by *Giardia* sp., *Cryptosporidium* sp., Norwalk-like virus, *Campylobacter* sp., *Salmonella* sp., *Escherichia coli* O157:H7, and *Shigella* sp. Distribution system deficiencies such as cross-connections, back-siphoning, improperly protected storage facilities, construction, and pipe repair, were the most common cause of outbreaks in community systems. Waterborne disease data from the United Kingdom, Scandinavia, Canada, Japan, Israel and elsewhere suggests similar etiological patterns (Hrudey and Hrudey, 2004) and highlight the importance of continued improvements in microbial monitoring for the water industry. Moreover, recent waterborne outbreaks have clearly demonstrated that consumption of contaminated drinking water is not solely a problem in the developing world; all of the industrialized nations of the world, with highly efficient and modernized treatment plants, continue to experience waterborne disease outbreaks. The causative agents of these waterborne disease outbreaks included protozoa, viruses, and bacteria, when an agent was identified. However, it is generally recognized that no agents are identified in approximately 50% of waterborne outbreaks.

## PROBLEMS RELATED TO THE PRESENCE OF SPECIFIC MICROORGANISMS

Protozoan parasites such as *Cryptosporidium parvum* and *Giardia duodenalis* have been recognized as important etiologic agents of waterborne disease outbreaks. *C. parvum* is of major concern because it exhibits relatively high resistance to chlorine-based disinfectants at the doses applied in treatment plants, has a low infectious dose, and no drug approved for prophylaxis or therapy is currently available. It can lead to severe illness and even death in severely immunocompromised individuals. There have been many documented outbreaks of cryptosporidiosis associated with either drinking or recreational water in the U.S. (Fayer et al., 2000). The largest outbreak occurred during 1993 in Milwaukee, Wisconsin, resulting in estimates of approximately 403,000 illnesses and 100 deaths (MacKenzie *et al.*, 1994), although some authors believe these figures are over estimations (Hunter and Syed, 2001). Despite everything that has been learned about *Cryptosporidium* by the surge in research precipitated by the Milwaukee incident, waterborne outbreaks still occur. For example, an outbreak of waterborne cryptosporidiosis occurred in Canada in 2001 and was estimated to have sickened up to 7,100 people (Hrudey and Hrudey, 2004).

*G. duodenalis* has historically been the most frequently identified cause of waterborne disease outbreak (Craun, 1990). The first recognized outbreak of giardiasis in the U.S. occurred in New York State in 1974 – 1975 with 350 confirmed cases (Shaw, 1977). From 1971 to 1985, 92 waterborne disease outbreaks were attributed to *Giardia*, with a total of approximately 24,000 infected individuals. Between 1976 and 1994, *Giardia* was the most frequently identified etiologic agent in outbreaks caused by drinking water (Kramer et al, 1996). During 1993 and 1994, *G. duodenalis* was the causative agent in 9 out of the 39 (23 percent) waterborne disease outbreaks in the U.S., where the etiological agent was identified (Kramer et al, 1996). These outbreaks were associated with both surface water and groundwater intended for drinking, and four outbreaks were associated with recreational activity (swimming in pools, rivers, and lakes). There were four waterborne giardiasis outbreaks in the U.S. during 1997 and 1998, totaling 159 confirmed cases all associated with drinking unfiltered surface water or untreated groundwater (Barwick et al., 2000). These incidents, as well as many others that occurred around the world, prompted the USEPA to regulate the level of *Giardia* cysts allowed in drinking water (USEPA, 1989).

Enteric viruses such as Hepatitis A virus, Norwalk virus, and Rotaviruses, are responsible for about 10% of the reported waterborne disease outbreaks where an etiologic agent was identified (Craun, 1992). It is likely that many of the outbreaks for which no etiologic agent was identified (~50% of all reported outbreaks) were caused by viruses because of the failure to look for them and the limitations of the current detection methodology. Therefore, enteric viruses are recognized as high priority emerging pathogens by the water industry. The Echoviruses, Coxsackieviruses and Adenoviruses are included on the USEPA Candidate Contaminant List (USEPA, 1998). The enteroviruses include 72 types and form a distinct group of enteric viruses; they consist of the polioviruses (3 types), coxsackie A (23 types) and B (6 types) viruses, echoviruses (32 types), and 5 types of the new enteroviruses, which are not incorporated into the poliovirus, coxsackievirus, or echovirus groups. These viruses can cause a variety of illnesses, including gastroenteritis, respiratory infections, conjunctivitis, paralytic conditions, herpangina, myocarditis, epidemic pleurodynia, and meningitis. The presence of human enteroviruses in water poses a public health risk because of their low infectious dose, fecal-oral route of transmission and their prolonged survival. The enterovirus group serves as the *de facto* indicator of human enteric virus contamination since they are almost always isolated in sewage, and cell culture assays are available which is not the case for many other enteric viruses.

*Escherichia coli* is a normal inhabitant of the human intestinal tract. Only a few strains of *E. coli* are capable of causing human illness, including diarrhea, by several different mechanisms. These enterovirulent *E. coli* have been implicated in several major waterborne outbreaks (Feng, 1995), including the *E. coli* O157:H7 occurrence in Walkerton, Ontario,

Canada, in the spring of 2000 during which 7 individuals died and more than 2,300 became ill (O'Connor, 2002). Five classes of *E. coli* implicated in diarrheal diseases have been recognized: enterotoxigenic (ETEC), enterohemorrhagic (EHEC), enteroinvasive (EIEC), enteropathogenic (EPEC), and enteroadherent-aggregative (EAEC). Unfortunately, strains of *E. coli* that cause disease are difficult to distinguish from those that are part of the normal human intestinal flora. Definitive identification requires the determination of the characteristic virulence factors associated with each group. Among the EHEC, the serotype O157:H7 is a rare variety of *E. coli* that produces large quantities of one or more related, potent cytotoxins that cause severe damage to the lining of the intestine. These Shiga-like toxins (SLT I and SLT II) are closely related to the toxin produced by *Shigella dysenteriae*. Importantly, the majority of *E. coli* O157:H7 strains identified in the U.S. do not exhibit β-glucuronidase activity (Monday et al., 2001) and so would not be recognized as *E. coli* by the standard 4-methylumbelliferyl-β-*D*-glucuronide (MUG)-based methods used by the water industry.

**Detection Methods**

A common factor with all these microorganisms is the lack of simple and rapid methods for their detection and identification in water. Schematics of selected detection and identification methods typically used in the water industry are presented in Figures 1.1 – 1.3. Detection methods for *C. parvum* and *G. duodenalis* in drinking water have several limitations. Following recovery and concentration of (oo)cysts using the currently approved technique (Method 1623; USEPA, 1999), the microorganisms are detected using an immunofluorescence assay (IFA) with confirmation through 4,6-diamidino-2-phenylindole (DAPI) staining and differential interference contrast microscopy. However, this technique recognizes a diversity of *Cryptosporidium* species because of the broad specificity of the primary antibody. Also, the method is costly, and requires extensive processing time.

Conventional methodologies for the detection of enteroviruses in water rely on a few established mammalian cell lines as hosts. The BGMK continuous cell line derived from African Green Monkey kidney cells is the most frequently used for the isolation of enteroviruses from environmental samples, because it provides high sensitivity to natural isolates and maintenance of the cells is relatively straightforward. However, there are a number of potential problems associated with the use of BGMK cells. Comparison of virus particle counts versus infectious units has demonstrated that the plaque-forming unit (PFU) represents only a fraction of the actual virus particles present. Between 100 and 1,000 virus particles per PFU of poliovirus 1 have been determined when measured in established cell lines (reviewed by De Leon and Sobsey, 1991). The particle count per infectious unit appears to be dependent on the cell line used for assay. Comparative studies with the human rhabdomyosarcoma (RD) and the PLC/PRF/5 hepatoma cell lines have indicated that the BGMK cells are more sensitive to polioviruses and coxsackie B viruses, but they are substantially less sensitive to echoviruses and coxsackie A viruses (De Leon and Sobsey, 1991). In addition, BGMK cells also support the growth of non-human enteroviruses including bovine enteroviruses 5 and 6 and porcine enterovirus 8 (De Leon, unpublished results). Finally, the currently approved cell culture method for total culturable enteric viruses utilizing BGMK cells, is labor intensive, costly and the results are not available for 4-6 weeks after onset of analysis.

Although *E. coli* is currently detected by conventional culturing techniques, it is recognized that these methods are considerably limited because of the presence of metabolically active but non-culturable bacteria (particularly when microorganisms have been injured by disinfectants), the limited number of samples that can be processed, the inability to identify the strain, and lengthy incubation periods. In addition, the culturing techniques used by the water industry (Colilert, EC-MUG [4-methylumbelliferyl-β-D-glucuronide], and others) may not detect some strains of *E. coli* O157:H7 due to the lack of a functional β-glucuronidase enzyme in some strains (Monday et al., 2001).

The drinking water industry needs simple and rapid assays for pathogen detection and identification in water. Currently approved or recommended methods are still based on lengthy conventional culturing or microscopic enumeration techniques. In addition, these methods are not designed for high throughput or rapidity of analysis, and the inclusion of a direct link to automation and robotic technology has seldom been a consideration. Microbial detection assays that incorporate efficiency of detection, the ability to process large sample numbers, and simplicity of operation are needed to serve the long-term needs of the water industry. Future methods need to incorporate specific detection of target microorganisms at the species and/or strain level, minimal or no risk of false identification, and sensitive detection to the low levels of pathogens typically encountered in natural and treated water.

Molecular techniques provide new avenues for detection and identification of specific microorganisms (Figure 1.4). Nucleic acid amplification techniques such as the polymerase chain reaction (PCR) can detect very low concentrations of microbial DNA or RNA; an enrichment of $10^5$-$10^6$ copies of target nucleic acid sequences can be obtained in a few hours *in vitro* (Ou *et al.*, 1988). Molecular assays, and in particular PCR, have been proposed and used for many years as alternative techniques for the detection of a wide range of pathogens in water (Cromeans et al., 2004; Jiang et al., 2001; Lang et al., 1994, Reynolds et al., 2001; Rochelle et al., 1997a; Rochelle, 2001; Schwab et al., 1995; Schwab et al., 1996; Stinear et al., 2001). Numerous studies have now demonstrated efficient amplification of DNA and RNA extracted directly from environmental samples, using PCR or other in-vitro amplification methods.

Following nucleic acid amplification, amplicons specific to the target microorganism can be confirmed by internal probe hybridization. During hybridization, a segment of radioactively or non-radioactively labeled cloned or synthesized nucleic acid (a probe) is hybridized to the complementary sequence of the amplified genome fragments. Such probe-based confirmation is an essential step when using molecular methods to detect pathogens in environmental samples because a variety of factors can interfere with the specificity of primer-based amplification reactions. These include poor primer design, cross reactivity with closely related species/strains, non-optimized reactions, and co-extraction of substances from environmental samples that may reduce primer specificity. Pathogen identification in environmental samples solely on the basis of amplicon length, as visualized by gel electrophoresis, may lead to false positive detection.

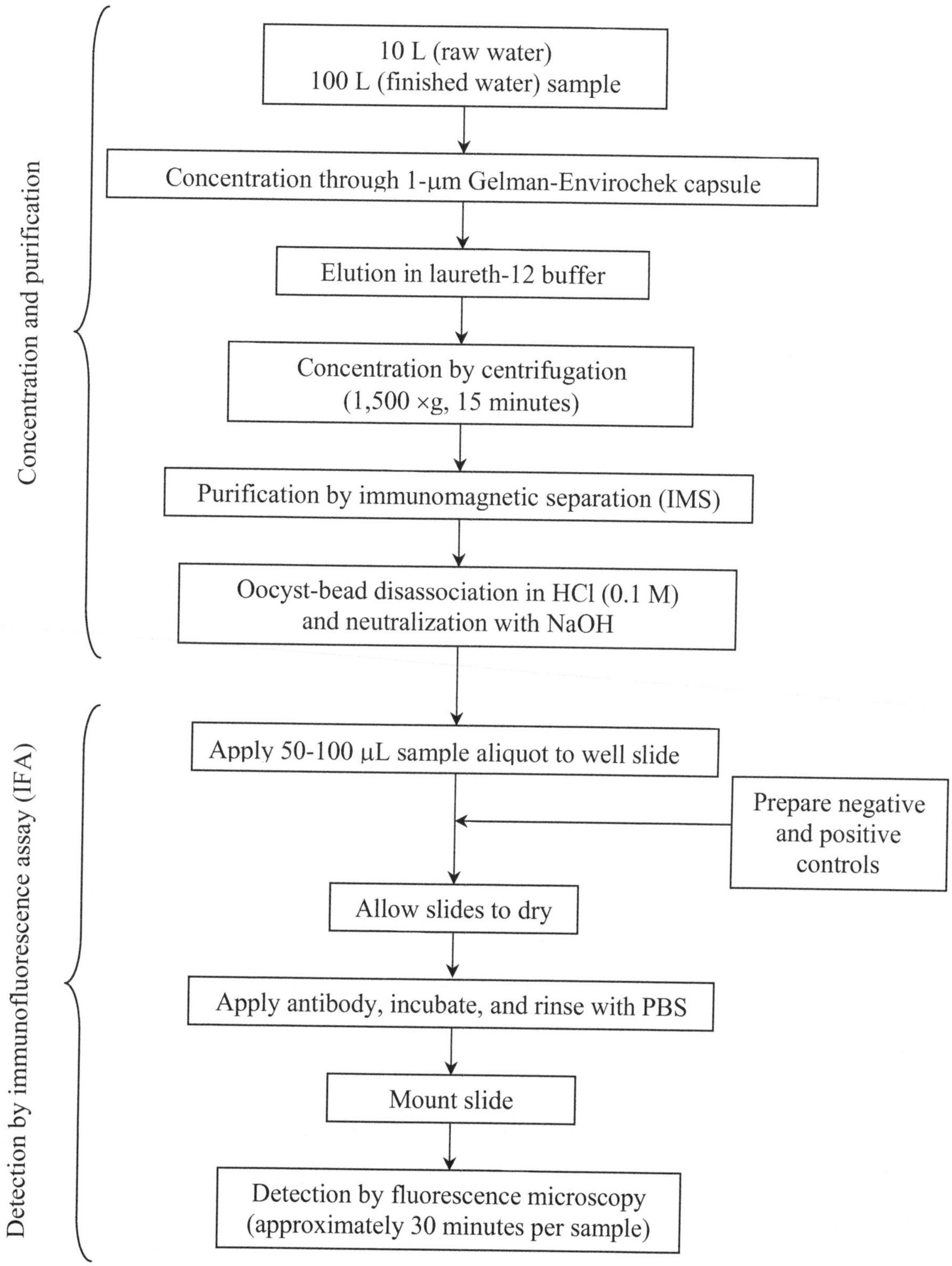


**Figure 1.1. Conventional detection method for *Cryptosporidium* and *Giardia* in water (Method 1623; USEPA, 1999).**

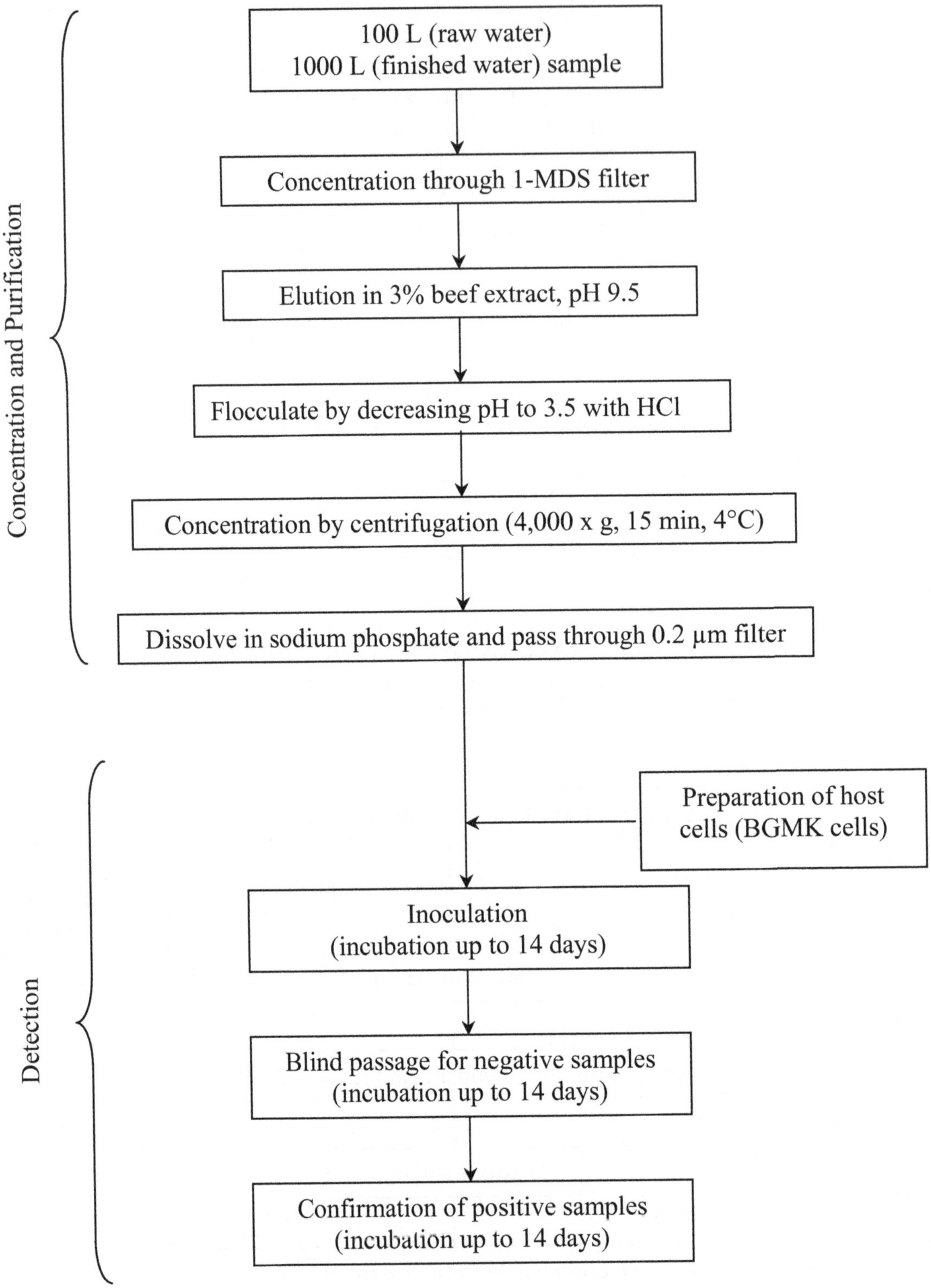


**Figure 1.2. Conventional detection method for enteric viruses in water (USEPA, 1996).**

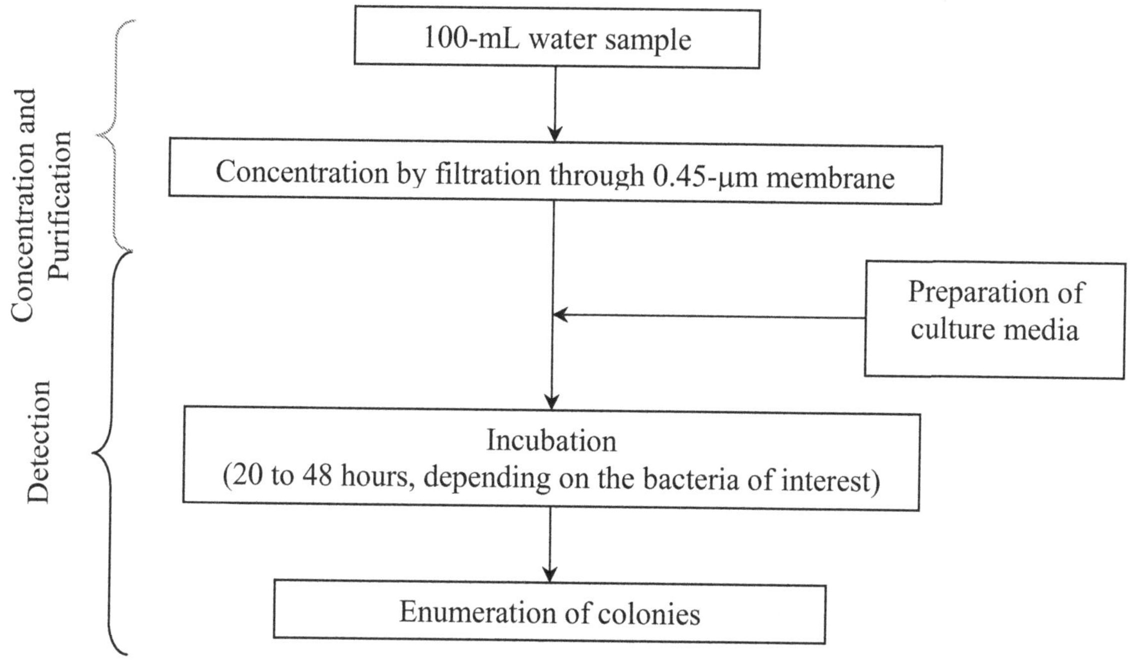


**Figure 1.3. Typical detection technique for bacteria in water.**

## PROPOSED ASSAY

The field of clinical diagnostics has developed fully automated robotic systems that process microbial samples from start-to-finish with minimal need for an operator. Results are delivered directly into a computer linked to the system. Most of these automated systems utilize a multi-well microplate and microplate reader format. This project developed a simple and rapid screening assay for microbial pathogens using molecular methods in a microplate format, similar to the clinical systems (Figure 1.5). The assay consists of nucleic acid amplification followed by hybridization in 96-well streptavidin-coated microplates with non-radioactively labeled probes and detecting the hybrids by colorimetric or fluorometric methods. With this approach, high sensitivity and specificity are achieved by nucleic acid amplification while confirmation by internal probe hybridization increases the specificity and avoids potential detection of false positives from sub-optimal amplification. In addition, microplate hybridization significantly decreases analysis time when compared to membrane hybridization (over 20 hours) or the conventional techniques used for pathogen detection (several days in some cases). Lengthy procedures for pathogen detection in water, particularly in finished drinking water, do not provide information in a timely manner to help prevent or minimize consumer exposure to these microbial contaminants. High throughput, rapid reading and a high potential for automation at low cost are provided by the use of multi-well microplates and a microplate reader format. Since the majority of water samples are expected to be negative for pathogens, a rapid screen assay allows the option of eliminating negative samples from further analysis, and conducting more lengthy and costly cultural assays for pathogen enumeration or infectivity determinations, on positive samples only. The proposed assay focused on the detection of microorganisms following their recovery from water samples.

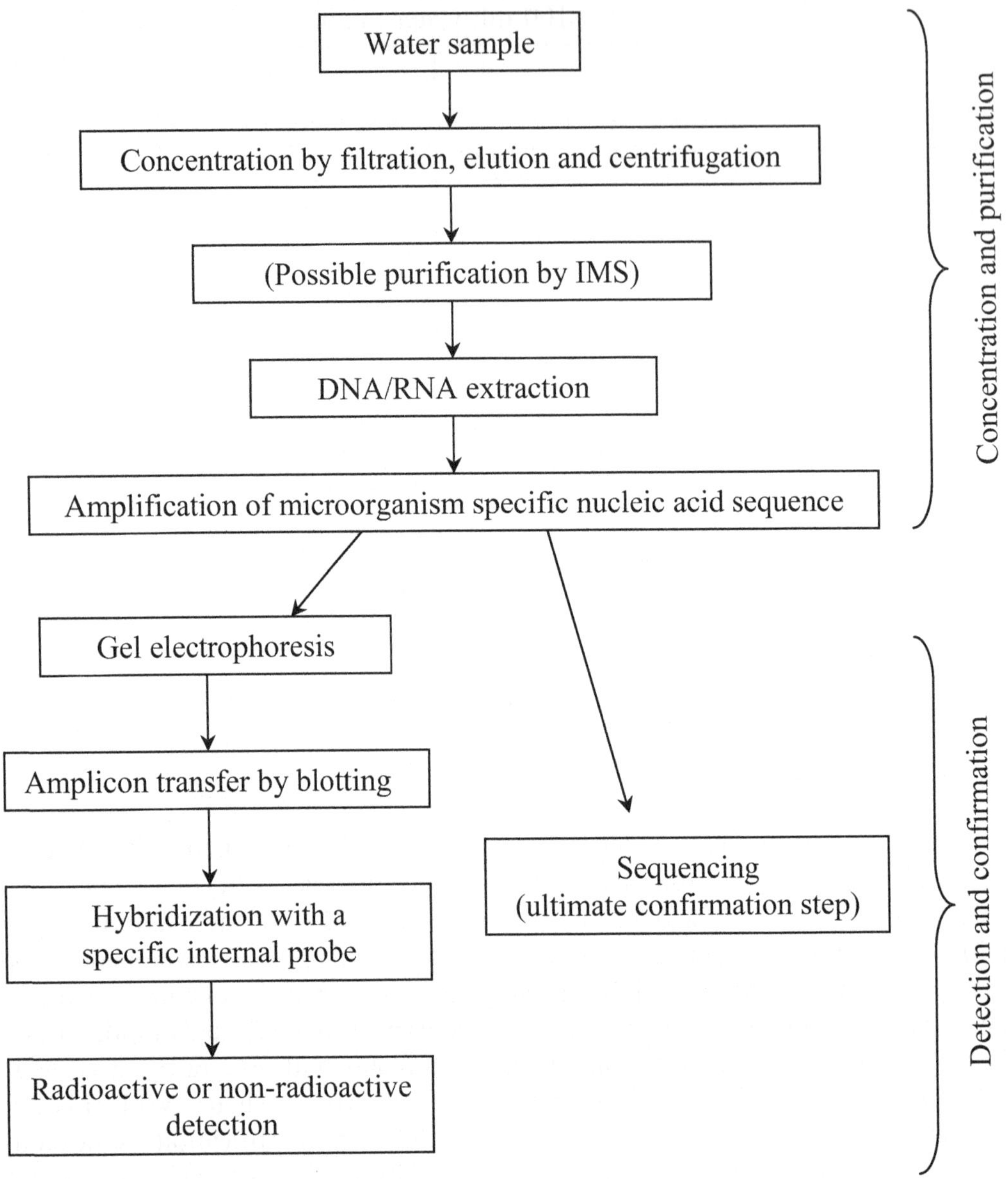


**Figure 1.4. Molecular methods as typically applied to environmental water samples.**

## OBJECTIVES

The overall objective of the project was to develop a sensitive, specific, robust, and rapid screening assay for the detection of pathogens in water, using commercially available supplies and reagents. The project aimed to develop a simple assay for rapid screening of selected pathogens in water: *C. parvum*, *G. duodenalis*, echovirus 1, and *E. coli* O157:H7. The specific objectives were:

1. Optimize a prototype assay comprising PCR amplification and microplate hybridization using *C. parvum* as a model organism.

2. Evaluate the adaptability of the assay to detect other organisms using echovirus 1, *G. duodenalis*, and *E. coli* O157:H7 as representatives of each type of organism (viruses, protozoa, and bacteria).
3. Further develop the assay by comparing the prototype detection assays for *C. parvum* and enteroviruses with:
    a. Other microplate hybridization approaches.
    b. Fluorescence-based detection.
    c. Peptide nucleic acid (PNA) probes instead of DNA probes.
4. Evaluate the assay using environmental water samples seeded with *C. parvum* and echovirus 1.

## SELECTION OF MICROORGANISMS

The rapid screening assay was evaluated using two protozoa, one bacterium, and a representative enteric virus to assess the adaptability of the assay to different microorganism groups. Ideally, the assay would be readily transferred for use with other microorganisms starting from optimized conventional PCR and hybridization conditions. The microorganisms selected were *C. parvum*, *G. duodenalis*, echovirus 1, and *E. coli* O157:H7 because of their significance to the water industry. They have been detected in source and/or drinking water, and have been identified as the causative agent in waterborne outbreaks. There is a lack of simple and rapid methods for their detection and identification, and they are currently regulated or are likely to be regulated in the near future.

## ORGANIZATION OF REPORT

Chapter 2 presents all materials and methods used to develop and validate the assay during this project. Chapter 3 addresses the results and discussion including comparison of the different approaches examined, sensitivity and specificity results, validation of the assay with seeded environmental water samples, as well as the evaluation of a kit developed through a collaborative effort between the investigators and a manufacturer. The study conclusions and recommendations for future work are presented in Chapter 4. Detailed assay protocols are provided as appendices.

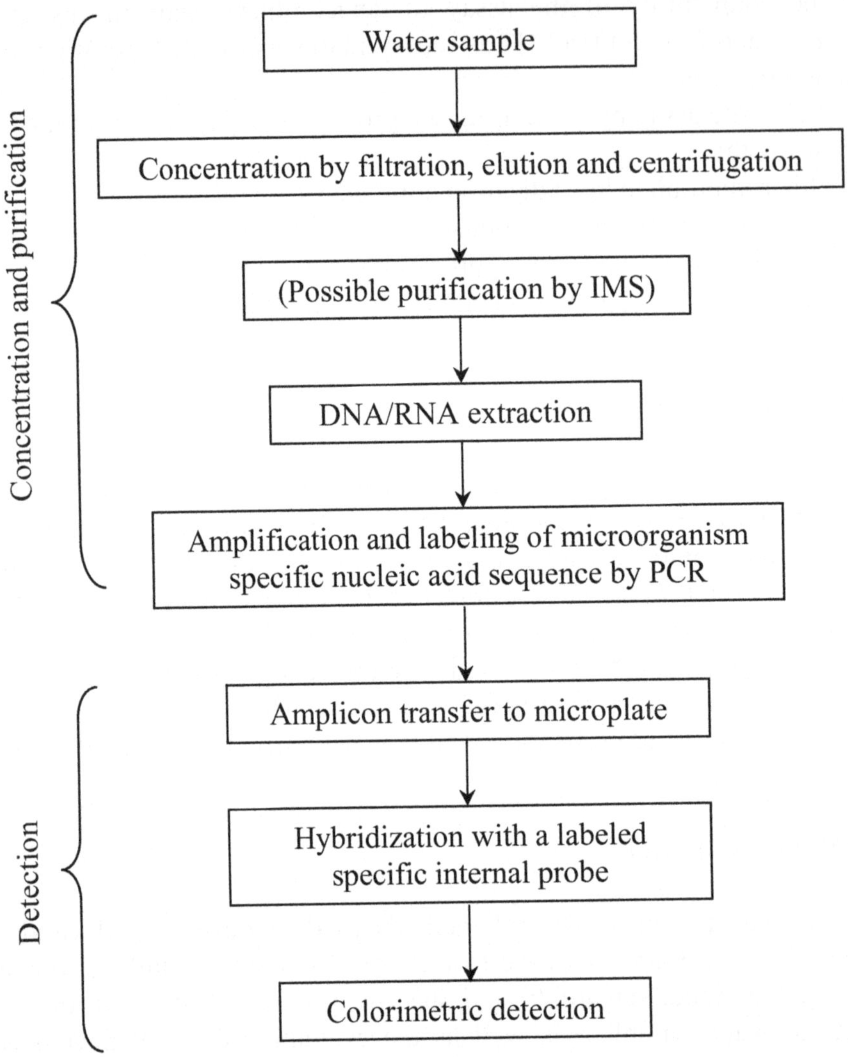


**Figure 1.5. Proposed approach for the detection of microorganisms using molecular methods in a microplate format.**

# CHAPTER 2
# MATERIALS AND METHODS

During this study, different PCR, microplate hybridization and detection approaches were examined. Figure 2.1 presents the amplification and hybridization approaches evaluated. For each of these approaches, PCR amplification of the DNA extracted from target microorganisms was conducted, followed by hybridization and signal detection in a 96-well streptavidin coated microplate. Different haptens were incorporated during the amplification reaction either to attach the amplicons to the microplate wall or detect the amplicon-probe hybrids, depending upon the approach used. The prototype approach (PCM/ML-dUTP, simplified to PCM), involved probe-coated microplates (PCM) and multiple DIG labels (ML) incorporated into the amplicons via DIG-dUTP during amplification. A detailed protocol of this approach is presented in Appendix A. The other main approach evaluated was named ACM/SL-PRB (or ACM), and involved microplate coating with the amplicons (ACM) and single detection label (SL) on the 5'-end of the probe (PRB). A detailed protocol of this approach is presented in Appendix B. Variations of these approaches were investigated and are presented throughout the report.

This chapter presents the optimum conditions and conditions that were common to a significant number of experiments. Other techniques that were used to develop the assay, determine the rate limiting step(s), and optimal reagent combinations are also presented. These experiments used microplates and a slot blot manifold with Nylon membranes. The use of membrane slot blots at specific steps during the process followed by appropriate detection techniques of bound products identified the limiting steps. Southern transfers of electrophoresis gels were also used to evaluate different amplification conditions. The technical details of the evaluation experiments conducted with water samples, and the quality assurance and quality controls performed during each experiment are also presented in this chapter.

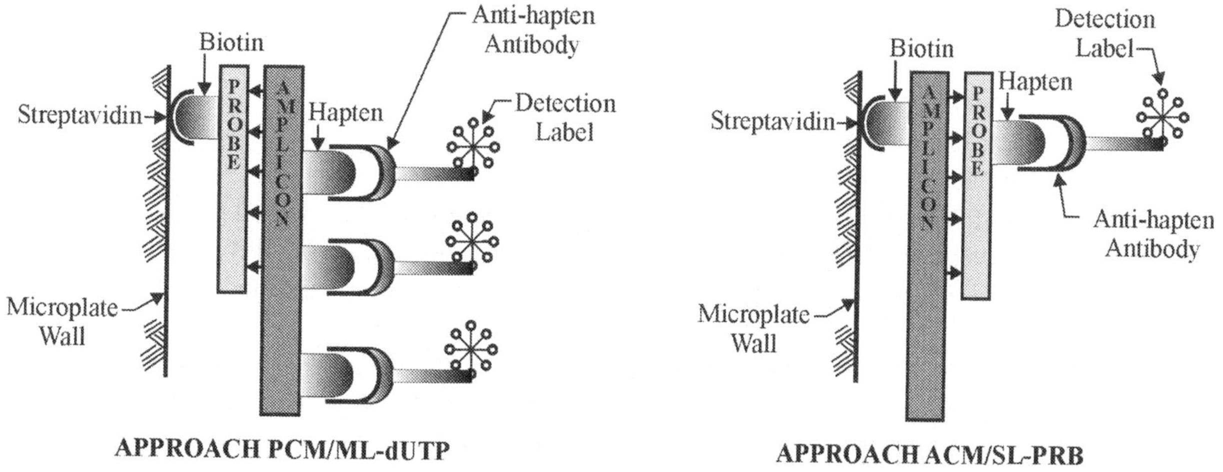


**Figure 2.1. Microplate hybridization and detection approaches for the rapid screening assay. PCM/ML-dUTP, Probe-coated microplate and amplicon containing multiple labels incorporated during PCR as digoxigenin-dUTP; ACM/SL-PRB, Amplicon-coated microplate with amplicon containing a single label incorporated as a 5'-labeled primer during PCR.**

## SOURCES, PREPARATION, AND HANDLING OF MICROORGANISMS

### *Cryptosporidium parvum*

Purified *C. parvum* oocysts of the Iowa isolate were purchased live in water from the University of Arizona (Sterling Parasitology Laboratory, Department of Veterinary Science, Tucson, Arizona). Oocysts of the TAMU (Okhuysen et al, 1999) and Moredun (Blewett et al, 1993) isolates were supplied to the Sterling Parasitology Laboratory by C. Chappell (University of Texas, Houston) and S. Wright (Moredun Research Institute, Peniculk, Scotland), respectively, and propagated in calves at the Sterling Parasitology Laboratory. Oocysts were stored in saline antibiotic solution at 4°C prior to utilization.

Oocyst concentrations were determined by hemocytometer enumeration (Bright-line; Hausser Scientific, Horsham, Pennsylvania) and lower concentrations were obtained by dilution of the stock. As per the hemacytometer manufacturer instructions, 5 small squares were counted on both sides of the hemacytometer (10 squares total), and the average was calculated. The average standard deviation was less than 20% between the 10 counts.

Serial dilutions of *C. parvum* oocysts were prepared in sterile phosphate buffered saline (1X PBS) or 1X Tris-EDTA (10 mM Tris, 1 mM ethylenediaminetetraacetic acid [EDTA], pH 7.4), and oocysts from each dilution were extracted in separate tubes. Over the course of the study, oocyst DNA was extracted using a variety of methods. A discussion about the implications of using different methods is presented in Chapter 3. The most complex extraction protocol was:

1. Oocyst suspensions were centrifuged (3,000xg, 5 min, 4°C) and re-suspended in 300 μL Tris-EDTA buffer (50 mM Tris, 20 mM EDTA, pH 7.5).
2. Following addition of 30 μL proteinase K (Roche Applied Science, Indianapolis, Indiana), suspensions were incubated at 37°C for 30 min.
3. After addition of 200 μL of 2% N-laurylsarcosine sodium salt (Sigma Chemical Corp., St. Louis, Missouri), tubes were incubated at 37°C for 15 min.
4. Addition of 175 μL 5 *M* sodium chloride (NaCl) and 170 μL of 5% hexadecyltrimethyl-ammonium bromide/NaCl (CTAB/NaCl).
5. Incubation at 65°C for 30 min.
6. Addition of 750 μL phenol/chloroform/isoamyl alcohol (25:24:1 v/v/v) (Sigma).
7. Extracts were centrifuged to pellet the debris and DNA in the supernatant was precipitated by addition of 330 μL of isopropanol (Sigma) and incubation at -20°C overnight.
8. DNA was recovered by centrifugation and washed with 500 μL of 70% ice-cold ethanol.
9. The final DNA pellet was dried in a desiccator (AS160 Automatic SpeedVac; Savant Instruments, Holbrook, New York) and re-suspended in 100 μL sterile molecular biology grade water (18 Megohm, Sigma).

For other experiments, oocysts were washed and resuspended in either sterile PBS or 10 mM Tris-HCl buffer (pH 7.5). DNA was then extracted from oocysts by heat lysis (95°C for 10 min) followed by six cycles of freezing (-70°C for 10 min) and thawing (65°C for 5 min), and recovery of DNA by centrifugation.

### *Giardia duodenalis*

*G. duodenalis* cysts of the WB strain were obtained from Hyperion Research Laboratory (Medicine Hat, Alberta, Canada) and the H3 isolate was obtained from Waterborne, Inc. (New Orleans, Louisiana). The cysts were stored at 4°C prior to extraction. The techniques described above for dilution and enumeration of *C. parvum* oocysts and extraction of DNA were also used for *G. duodenalis* cysts.

### Enteroviruses

Echovirus 1 (VR-1038), obtained from the American Type Culture Collection (ATCC; Manassas, Virginia), was used as representative of the enterovirus group. It was propagated in BGMK cells grown in 175-cm$^2$ flasks at a multiplicity of infection (MOI) of 0.01-0.1 plaque-forming units (PFU) per cell. When cytophatic effects (CPE) reached 90%, the cells were frozen and thawed three times. The cell supernatants were harvested, filtered through 0.22-μm porosity Acrodisc membranes (Pall Gelman, East Hills, New York), and stored at -80°C. Immediately prior to reverse transcription, RNA was released from the enteroviruses by heating the samples at 80°C for 5 min and then snap-cooled on crushed ice.

Echovirus 1 was quantified using a plaque forming method on BGMK cell monolayers in 6-well microplates, with replicates for each virus dilution (Melnick *et al.*, 1979). Water samples were assayed in 25-cm$^2$ flasks by the quantal CPE assay on BGMK cell monolayers following the Information Collection Rule (ICR) method guidelines (USEPA, 1996).

### *Escherichia coli* O157:H7

The *E. coli* O157:H7 strain used during this study was isolated from a patient in Cabool, Missouri (EPA #932), and received from Dr. Anne Camper, Montana State University (Bozeman, Montana). Its identity was confirmed by PCR with primers targeting the shiga-like toxin 2 gene and a gene necessary for O-antigen biosynthesis, and serological testing by latex agglutination. It was maintained on R2A plates at 37°C. A colony was picked, streaked on fresh tryptic soy agar plate and incubated overnight at 37°C. Colonies were then picked, resuspended in tryptic soy broth (TSB), and incubated overnight with agitation (125 rpm) at 37°C. The suspension was washed twice by centrifugation at 4°C for 10 min and re-suspended in 5 mL phosphate buffer (Method 9050 C 1.a; APHA, 1998). Serial dilutions were prepared and an aliquot of each was used for DNA extraction using the method detailed above while the rest was used for enumeration on m-Endo medium following Method 9222 B (APHA, 1998).

### Microorganisms Used for Evaluating Specificity

Selected microorganisms that may be found in source or treated waters, or may interfere with the detection of the target organism(s), were used to evaluate assay specificity (Table 2.1). Human DNA was also used during specificity evaluation. Bacteria were cultured in appropriate growth media and colonies were picked and resuspended in 50 mM Tris/20 mM EDTA. Microorganism suspensions were filtered on 0.22-μm polycarbonate membranes and DNA was extracted directly from the membrane using the method detailed above. Viruses were recovered from cell culture and the RNA was released by heat lysis of the host cells.

### DNA Assessment

For all organisms used in this study, the suitability of the purified DNA for amplification was pre-tested by PCR with unlabeled primers and reagents. Following PCR, amplicons were observed by gel electrophoresis on agarose or polyacrylamide gels stained with ethidium bromide, and photographed on an ultraviolet transilluminator (Sambrook *et al.*, 1989). DNA was stored in water at -20°C for further utilization. In the case of enteroviruses, the suitability of the purified RNA was pre-tested by reverse transcription and then PCR using unlabeled primers and reagents, and the RNA was stored at -80°C until further utilization.

## NUCLEIC ACID AMPLIFICATION

PCR was performed on DNA extracts, or from RNA previously converted to a DNA copy (cDNA) by the reverse transcriptase reaction. All primers were previously published or tested (Table 2.2). Optimization of amplification conditions was performed for *C. parvum*, the model microorganism, during the development of the prototype method, and is presented in Chapter 3.

Amplification reactions were performed in 1X PCR buffer containing 10 mM Tris-HCl (pH 8.3), 50 mM potassium chloride (KCl) and 0.001% (w/v) gelatin, along with 0.01 μg/μL bovine serum albumin (BSA) and 0.025 U/μL of polymerase enzyme. The concentrations of magnesium chloride ($MgCl_2$), dNTPs, primers, polymerase enzyme, and thermal cycling activity and annealing temperatures were dependent on the primers and approach tested, and are explained in Chapter 3. For Approach PCM/ML-dUTP, the optimum concentrations of dATP, dCTP and dGTP were 100 μM with 200 μM of dUTP (including labeled and unlabeled dUTP). The primer concentration was 500 nM for Approach PCM/ML-dUTP, and 250 nM for Approach ACM/SL-PRB. Various polymerase enzymes were evaluated including AmpliTaq and AmpliTaq Gold (Applied Biosystems, Foster City, California), Platinum Taq (Invitrogen, Carlsbad, California), Tth and Expand High Fidelity polymerases (Roche Molecular Biochemicals). The $MgCl_2$ concentrations and polymerase enzymes that were selected to obtain optimum sensitivity and specificity are presented in Table 2.3, and the specific conditions for Echovirus 1 are presented in a separate section below.

**Table 2.1**
**Organisms used for evaluation of assay specificity**

| Organisms | Source[a] |
|---|---|
| **Bacteria:** | |
| *Aeromonas hydrophila* | Natural water isolate (MWD) |
| *Escherichia coli* (XL-1 Blue) | Stratagene, La Jolla, California |
| *Klebsiella pneumoniae ozaenae* | ATCC 11296 |
| *K. pneumoniae pneumoniae* (type 2) | ATCC 9997 |
| *K. pneumoniae pneumoniae* (type 3) | ATCC 13883 |
| *K. pneumoniae pneumoniae* (type 30) | ATCC 27736 |
| *Klebsiella trevisanii* | ATCC 33558 |
| *Klebsiella oxytoca* | ATCC 43863 |
| *Salmonella typhimurium* | ATCC 14028 |
| *Clostridium perfringens* | ATCC 13124 |
| *Yersinia enterocolitica* | ATCC 23715 |
| *Shigella sonnei* | ATCC 25931 |
| *Enterobacter cloacae* | ATCC 13047 |
| *Citrobacter freundii* | ATCC 8090 |
| *Pseudomonas fluorescens* | ATCC 13525 |
| *Nitrosomonas europea* | ATCC 19718 and ATCC 25978 |
| *Comamonas testosteroni* | ATCC 11996 |
| **Mammalian Cells:** | |
| BGMK | USEPA |
| Caco-2 | ATCC HTB-37 |
| HCT-8 | ATCC CCL-244 |
| **Protozoa:** | |
| *Cryptosporidium muris* | Waterborne, Inc., New Orleans, Lousiana |
| **Viruses:** | |
| Rotavirus | Charles C. Gerba, University of Arizona |
| Hepatitis A | Mark D. Sobsey, University of North Carolina |
| Bovine enteroviruses 1 to 7 | CDHS-VRDL |
| Porcine enteroviruses 1, 3 – 8, 8a, 8b, 8c | CDHS-VRDL |
| Poliovirus 1 | CDHS-VRDL |
| Coxsackie virus strains A9, A12, B3, B5 | CDHS-VRDL |
| Echovirus 1 | ATCC VR-1037 |
| Echovirus 7 | CDHS-VRDL |
| **Yeast:** | |
| *Saccharomyces cerevisiae* | Sigma Chemical Corp. |

[a] MWD, Metropolitan Water District of Southern California; ATCC, American Type Culture Collection; CDHS-VRDL, California Department of Health Services - Viral and Rickettsial Disease Laboratory.

**Table 2.2**
**PCR primer description**

| Organisms | Primer pairs[a] | Product size | Sequences[b] (5' – 3') | $T_a$[c] | References |
|---|---|---|---|---|---|
| *C. parvum* | CPHSP2423F<br>CPHSP2764R | 361 bp | AAA TGG TGA GCA ATC CTC TG<br>CTT GCT GCT CTT ACC AGT AC | 55°C | Rochelle et al., 1997b |
| | CPHSP2465F<br>CPHSP2764R | 319 bp | GGA TGT TGC TCC ATT ATC AC<br>CTT GCT GCT CTT ACC AGT AC | 52°C | F- this project<br>R- Rochelle et al., 1997b |
| *G. duodenalis* | ABB-97F<br>ABB-220R | 163 bp | AGG GCT CCG GCA TAA CTT TCC<br>GTA TCT GTG ACC CGT CCG AG | 55°C | Abbaszadegan et al., 1997 |
| *E. coli* | SLT-II-F<br>SLT-II-R | 482 bp | CTT CGG TAT CCT ATT CCC GG<br>GGA TGC ATC TCT GGT CAT TG | 60°C | Olsvik and Strockbine, 1993 |
| Human enteroviruses | PanEV-F<br>PanEV-R | 197 bp | CCT CCG GCC CCT GAA TGC<br>(A/T)CA C(T/C)G G(A/G)T GGC CAA TCC AA | 57°C | De Leon et al., 1990 (modified) |

[a] F: Forward primer; R: Reverse primer
[b] (A/T), (T/C), (A/G): 50/50 mixture of indicated bases.
[c] Annealing temperature.

**Table 2.3**
**Selected $MgCl_2$ concentration and polymerase enzymes for PCR amplification**

| Organisms | $MgCl_2$ concentration | Polymerase enzyme | |
|---|---|---|---|
| | | Approach PCM/ML-dUTP | Approach ACM/SL-PRB |
| *C. parvum* | 1.5 mM | Platinum Taq | AmpliTaq Gold |
| *G. duodenalis* | 2.5 mM | AmpliTaq Gold | AmpliTaq Gold |
| *E. coli* O157:H7 | 3.0 mM | AmpliTaq Gold | AmpliTaq Gold |

Platinum Taq: Invitrogen, Carlsbad, California.
AmpliTaq Gold: Applied Biosystems, Foster City, California.

dUTP was substituted for dTTP in the amplification reactions so that uracil DNA-glycosylase (UNG, 1 U per 100-μL reaction; Applied Biosystems) could be included to prevent contamination by amplified products from previous amplification reactions. Since UNG is active up to 50°C, it decreases the formation of primer artifacts. The incorporation of dUTP into amplicons does not interfere with subsequent probing of the product. Therefore, prior to PCR activity, all samples were incubated at 24°C for 10 min to allow UNG activity, then at 95°C for 10 min to destroy UNG and activate the polymerase. For the PCM approach requiring labeled amplicons, amplification reactions contained DIG-11-dUTP (Roche Applied Science) in a 1:4 ratio of labeled:unlabeled dUTP (total concentration of 200 μM). The primers were unlabeled for Approach PCM/ML-dUTP.

For Approach ACM/SL-PRB, both primers were biotin labeled via a 15-carbon chain link (BIOTEG). Initial experiments utilized an unlabeled forward primer (CPHSP2423F) and biotin-labeled reverse primer (CPHSP2764R) for *C. parvum* detection. Subsequent experiments demonstrated that improved sensitivity was achieved when both primers were biotin labeled. Following optimization, a standardized thermal cycling profile was used for each test organism (Table 2.4).

**Table 2.4**
**Thermal cycling conditions**

| Organisms | Denaturation | Annealing | Extension | Final extension |
|---|---|---|---|---|
| *C. parvum* | 94°C for 1 min | 55°C for 1 min | 72°C for 2 min | 72°C for 10 min |
| *G. duodenalis* | 94°C for 1 min | 55°C for 1 min | 72°C for 2 min | 72°C for 5 min |
| *E. coli* O157:H7 | 94°C for 45 sec | 60°C for 45 sec | 72°C for 90 sec | 72°C for 10 min |
| Echovirus 1 | 94°C for 30 sec | 57°C for 30 sec | 72°C for 1 min | 72°C for 7 min |

The use of double or semi-nested PCRs for *C. parvum* detection using the PCM approach was investigated. The first PCR included UNG to eliminate amplicons originating from previous PCR, while the second reaction did not include UNG to avoid destruction of the amplicons obtained during the first amplification (Table 2.5). The first PCR involved non-labeled dUTP, and the DIG label was incorporated in the second reaction. The double PCR experiments were conducted with primer set CPHSP2423F and CPHSP2764R. For the semi-nested PCR experiments, the first reaction was performed with the CPHSP2423F and CPHSP2764R primer set, and the second reaction was conducted with CPHSP2465F and CPHSP2764R. Similarly, the use of double PCR was investigated for *C. parvum* detection using the ACM approach (CPHSP2423F and CPHSP2764R primer set for both reactions). The first PCR was unlabeled, and the biotin-labeled primer(s) were incorporated in the second reaction. Further method development demonstrated that double or semi-nested PCR was not necessary to achieve sufficient sensitivity. To verify amplification efficiency, amplicons were separated by agarose gel electrophoresis, stained with ethidium bromide, and photographed (Sambrook *et al.*, 1989).

**Reverse Transcription (RT)-PCR for Enteroviruses**

Enteroviruses were amplified by reverse transcription followed by PCR using the protocol of De Leon *et al.* (1990) with the commercially available reagent kits (Applied Biosystems). The reverse transcriptase mix contained the following constituents: 1X PCR buffer described above, 3 mM of $MgCl_2$; 1.2 mM of each dATP, dCTP, dGTP and dTTP; 0.63 U/mL of RNase Inhibitor; 1 U/mL of MuLV Reverse Transcriptase enzyme; and 4.0 μM random hexamers. To release the RNA from the enteroviruses, 5-μL sample aliquots were incubated at 80°C for 5 min with all reverse transcriptase reagents except the RNase Inhibitor and the MuLV Reverse Transcritase enzyme, and then cooled on crushed ice for 2 min. The RNase Inhibitor and the MuLV Reverse Transcriptase enzyme were added to complete the reaction volume to 32 μL, and the reverse transcriptase activity was performed: 15 min at 25°C, followed by 1 h at 42°C, then 5 min at 99°C.

The cDNA obtained was then amplified by PCR in the 1X PCR buffer described above, along with 3 mM $MgCl_2$ solution, 0.6 μM of each primer, 0.01 U/μL UNG and 0.025 U/μL AmpliTaq Gold polymerase enzyme. The concentrations of dNTPs varied with the approach tested: for Approach PCM/ML-dUTP, 250 μM of each dATP, dCTP and dGTP was used with 500 μM overall dUTP (labeled and unlabeled), whereas 500 μM of each dNTP was used for Approach ACM/SL-PRB. Following UNG activation (10 min at 24°C), and UNG denaturation and Taq activation (10 min at 95°C), the PCR activity was conducted according to the conditions presented in Table 2.4.

**Table 2.5**
**PCR conditions during double or semi-nested PCR**

| | UNG | dUTP |
|---|---|---|
| First PCR | + | Not labeled |
| Second PCR | – | DIG-11-dUTP |

## MICROPLATE HYBRIDIZATION

Following amplification, the presence of the target organism was confirmed by hybridization directly in microplates. Detailed laboratory protocols for Approaches PCM/ML-dUTP and ACM/SL-PRB are presented in Appendices A and B, respectively, and schematics of both approaches are presented in Figures 2.2 and 2.3. Samples were heat denatured at 95°C for 5 min, then at 99°C for 1 min, and a sample aliquot was transferred into 96-well microplates coated with streptavidin.

All DNA probes used were previously published or tested and are described in Table 2.6. DNA probes were hybridized at high stringency (dissociation temperature, $T_d$, -5°C) to ensure specificity. For Approach PCM/ML-dUTP, the probes were biotin labeled via a 15-carbon chain link (BIOTEG). All probes used in Approach ACM/SL-PRB were digoxigenin or fluorescein labeled to allow detection of the amplicon-probe hybrids.

Microplate hybridization was performed for 2 –3 h in a solution of 6X saline-sodium citrate (SSC; 1X SSC contains 0.15 M NaCl and 0.015 M sodium citrate), 1% blocking reagent (Roche Applied Science) and 0.02% (w/v) lauryl dodecasulfate sodium salt (SDS). The probe concentrations varied depending on the organisms and approaches tested, and are presented in Chapter 3.

Following hybridization, microplate wells were washed twice with a solution of 5X SSC and 20 mM Tris-HCl (pH 7.5) for 10 min at the appropriate temperature (Table 2.6.).

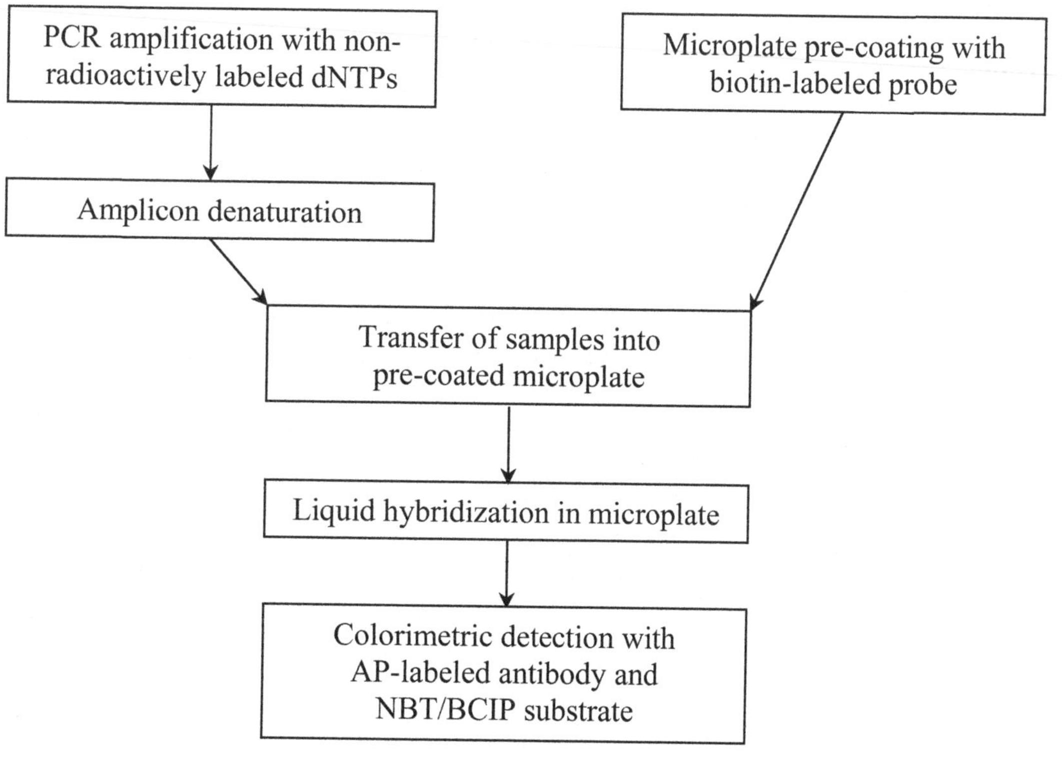


**Figure 2.2. Schematic of the probe-coated microplate approach (PCM/ML-dUTP).**

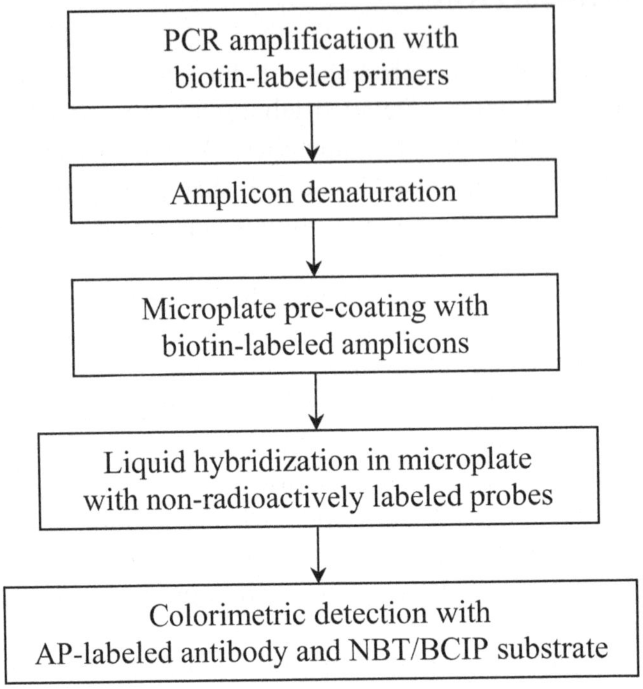


**Figure 2.3. Schematic of the amplicon-coated microplate approach (ACM/SL-PRB).**

**PNA Probe**

The PNA probes used during this study were a generous gift from Boston Probes, Inc. (Bedford, Maine; since acquired by Applied Biosystems). The most promising condition (*C. parvum* and Approach ACM/SL-PRB) was selected as a model to investigate the PNA probe. Therefore, the PNA probe was synthesized with a detection hapten (fluorescein) on the 5'-end. The sequence was a truncated form of the *C. parvum* CPHSP2475 DNA probe: CACTCGGTTTAGA. Optimum microplate hybridization conditions consisted of a 30 min incubation at 37°C in a solution of 100 mM Tris and 100 mM NaCl at pH 9.5.

**Table 2.6**
**DNA probe description**

| Organisms | Probes | Sequences (from 5' to 3') | Hybridization temperature /Washing temperature (°C) | References |
|---|---|---|---|---|
| *C. parvum* | CPHSP2475 | CCA TTA TCA CTC GGT TTA GA | 59°C / 57°C | Rochelle et al. (1997b) |
| | CPHSP2585 | CCA GAG TGG TGT CTT GAT C | 59°C / 57°C | This project |
| *G. duodenalis* | GLHSP184 | CAG GCC TTG GCG TTC CCG AAG | 65°C / 63°C | Rochelle et al. (1997a) |
| *E. coli* | SLT-IIP | CAC CGG GCA GTT ATT TTG C | 60°C / 58°C | Designed for this project |
| Human enteroviruses | PanEV-1 | GCG GAA CCG ACT ACT TTG | 55°C / 53°C | De Leon et al. (1990) |

## END-POINT COLORIMETRIC DETECTION

Following microplate hybridization and subsequent microplate washing, colorimetric detection was performed using an anti-hapten antibody conjugated to a reporter moiety (anti-DIG-alkaline phosphatase [AP], anti-DIG-fluorescein or anti-fluorescein-AP; Roche Applied Science). These antibodies were diluted in sterile detection buffer (100 mM Tris-HCl, 50 mM $MgCl_2$, 100 mM NaCl, pH 9.5) containing 10 µg/mL (1% w/v) BSA. The antibody solution was left in the microplate wells for 30 min at room temperature under gentle agitation. The microplate wells were washed twice with sterile detection buffer containing 10 µg/mL BSA, and each wash involved a 5 min incubation at room temperature. The colorimetric substrate consisted of 20 µL/mL of a solution of nitroblue tetrazolium salt and 5-bromo-4-chloro-3-indolylphosphate (NBT/BCIP; Roche Applied Science) in sterile detection buffer without BSA. The enzymatic reaction catalyzed by alkaline phosphatase on NBT/BCIP produces a purple precipitate that is visible under normal illumination. The optimum incubation period in NBT/BCIP for maximum sensitivity was 2 hours. Results were read at the end of the incubation period in NBT/BCIP, as well as during the two subsequent 5-min washes in sterile 1X Tris-EDTA buffer. The results were characterized using the following qualitative scoring scheme:

| | |
|---|---|
| - | no signal |
| +/- | weak, unclear signal |
| + | pale but distinct signal |
| ++ | medium signal intensity |
| +++ | strong signal |
| ++++ | very strong signal |

The colorimetric or fluorescent signal intensity was also read using a SpectraFluor Plus microplate reader (Tecan, Inc.; Research Triangle Park, North Carolina) with a high-energy Xenon flash lamp. For the absorbance reading, the wavelength selected was 550 nm (close to the mid-visible range to better mirror the visual inspection) with a resolution of 0.0001 OD. The fluorescence detection was conducted with excitation and emission wavelengths of 495 nm and 520 nm, respectively, with a sensitivity of ≤1.5 pg (≤4 fmole) fluorescein per well.

Following reading, all microplates were air dried, wrapped in plastic film and aluminum foil, and archived at 4°C. The signal is stable for many years when microplates are stored in this way. Detailed laboratory protocols for Approaches PCM/ML-dUTP and ACM/SL-PRB are presented in Appendices A and B.

## MICROPLATES

Streptavidin-coated microplates were selected for the development of this assay because they were readily available and they allow the use of the strong streptavidin-biotin bonding to capture either biotin-labeled probes (PCM approach) or amplicons (ACM approach). One molecule of streptavidin binds four molecules of biotin with an affinity constant ($K_d$) of $10^{-15}$ M, which is much higher than the affinity constant of most antibody-antigen bonds ($10^{-4}$ to $10^{-10}$ M; Gerhardt et al., 1994).

Initial experiments for *C. parvum* detection were conducted with 96-well streptavidin-coated, polystyrene microtiter plates (300-µL wells) or strips of eight 200-µL PCR tubes (Roche Applied Science). These standard plates and tubes were compared with StreptaWell High-Bind 96-well microplates (Roche Applied Science) that possess a much higher biotin binding capacity: 25 ng/well compared to 5 ng/well with regular microplates. Because results obtained with Approach PCM/ML-dUTP demonstrated better sensitivity with the High-Bind microplates than with the regular plates, these High-Bind microplates were used for the rest of the project, after verifying that the trends observed with both microplates were similar.

Prior to utilization, microplates were washed for 5 min with a solution of 50 mM Tris-HCl (pH 7.5) and 0.02% SDS in sterile deionized water under gentle agitation, to remove any manufacturing materials and residues that may potentially interfere with the assay. Following washing, the microplates were rinsed twice with sterile Milli-Q water.

## SLOT BLOT

Additional experiments were conducted using a slot blot apparatus to determine the rate limiting step(s) and/or optimal reagent combinations for hybridization. Prior to these experiments, the apparatus was cleaned in a solution of 0.1% SDS for at least 1 h, then rinsed in de-ionized water and air-dried. The Nylon membrane (Roche Applied Science) was soaked for at least 30 min in a 10X SSC solution. Two sheets of 3M paper (Whatman PLC; Kent, UK) were wetted in 10X SSC, placed on the slot blot apparatus, and overlaid with the pre-wetted Nylon membrane. Prior to sample filtration through the slot blot apparatus, 100 to 300 µL of 10X SSC was introduced in each well and filtered through the membrane. The same volume of 10X SSC was applied twice following sample filtration. The DNA-containing membrane was UV cross-linked using a Stratalinker 1800 (Statagene; La Jolla, California) and air-dried. The membrane was wrapped in a plastic film (Saran wrap) and kept at 4°C until further processing.

This common molecular technique was used for multiple purposes. As part of the quality control program it was used to verify the amplicons ability to hybridize to a probe or *vice versa* (with or without labels), and the presence of a specific hapten on amplicons or a probe. During assay troubleshooting, it was used to determine the potential release of amplicons or probe from the microplate well at specific steps of the assay. The detection techniques used varied with the objective of the experiments and were limited to a simple addition of an antibody and detection substrate, or complete hybridization followed by antibody and detection substrate addition. In most instances, a chemiluminescent detection was used with CDP Star substrate (Roche Applied Science).

## SOUTHERN TRANSFERS

Southern transfers of electrophoresis gels onto Nylon membranes were conducted to visualize the amplicons obtained from PCR amplification. DNA contained in the gel was denatured in a solution of 0.4 M sodium hydroxide (NaOH) and 0.6 M NaCl for 30 min, neutralized in 1.5 M NaCl and 0.5 M Tris-HCl for another 30 min, and equilibrated in 10X SSC for 10 min. A piece of Nylon membrane was soaked in a solution of 10X SSC for 30 minutes, then placed over the Nylon membrane, and isolated using plastic pellicles. Sheets of 3M paper were layered over the membrane, topped with a 500-g weight, and incubated for 14 –16 h at room temperature. Following transfer, the DNA was denatured in 0.4 M NaOH for 30 sec, and neutralized in a solution of 2X SSC and 0.20 M Tris for 10 min. The DNA-containing membrane was UV cross-linked and air-dried. The membrane was wrapped in a plastic pellicle and kept at 4°C until further processing.

Amplicons bound to the membrane were detected with or without hybridization, depending on the objective of the experiment. When hybridization was performed, the membrane was incubated under agitation at hybridization temperature for 90 minutes with SSC, blocking reagent, SDS and sarcosyl (the concentration of each reagent depended on the organism targeted). The buffer was then replaced with fresh buffer containing the probe, and incubated overnight at hybridization temperature under agitation.

This common molecular technique was used for multiple purposes, and the detection techniques used varied with the objective of the experiments. In some cases, amplicons were simply detected using an antibody and detection substrate, whereas at other times, a complete hybridization followed by antibody and detection substrate addition was performed. Most of the times, a chemiluminescent detection was used with CDP Star substrate.

## EVALUATION OF THE RAPID SCREENING ASSAY WITH SEEDED ENVIRONMENTAL WATER SAMPLES

To evaluate the applicability of the rapid screening assay for detection of microorganisms in natural waters, *C. parvum* and echovirus 1 were seeded into environmental water samples or water concentrates, recovered and concentrated using conventional techniques, and detected using the rapid screening assay and conventional methods. *C. parvum* oocysts were spiked into untreated influent water from Metropolitan's F.E. Weymouth Filtration Plant (La Verne, California). This plant receives a blend of California State Project water (SPW) imported from the Sacramento/San Joaquin delta in northern California, and Colorado River Water (CRW). The background level of *C. parvum* was not measured during these particular experiments but the running annual average in Weymouth plant influent is 0.003 oocysts/L based on in-house monthly monitoring using Method 1623 (USEPA, 1999). *Cryptosporidium* was detected only once in 2 years of monitoring at a concentration of 1 oocyst in 10 L.

Two aliquots of approximately 250 oocysts of *C. parvum* were seeded into two 10-L carboys. The oocysts from each carboy were recovered by Method 1623: the water was filtered through Gelman-Envirochek capsules (Pall Corporation – Life Sciences; Ann Arbor, Michigan), eluted in distilled water and a surfactant, concentrated by centrifugation, immuno-magnetic separation (IMS), and bead dissociation using 0.1 M HCl and 1.0 N NaOH for neutralization. It was important to submit the oocysts to the bead-dissociation step following the IMS as the acid involved might impair oocyst detection. The concentrates obtained were resuspended into sterile deionized water, for a total volume of 500 μL. From each carboy, the suspensions obtained were separated in three fractions, one for IFA (aliquots of 100 μL), and two for DNA extraction and analysis by the rapid screening assay using Approach ACM/SL-PRB and colorimetric detection. The two fractions used for the rapid screening assay consisted of different volumes (100 and 200 μL), leading to two different oocyst concentrations. Based on IFA enumeration of parallel fractions, the 100 μL and 200 μL aliquots contained the equivalent of 30 and 59 oocysts, respectively. Positive and negative controls consisted of formalized *C. parvum* oocysts provided with the antibody and sterile PBS, respectively. The basic steps of the IFA procedure as performed at Metropolitan are as follows:

- Apply 50 to 100 µL sample to the well slide, and spread over the well area. Allow the slide to dry (overnight at room temperature).
- Apply one drop of antibody over the entire well area. Crypt-a-Glo (Waterborne, Inc., New Orleans, Louisiana) is the direct antibody selected for these experiments. Incubate the slide at room temperature or 37°C in darkness for 30 min. Remove the antibody by absorbing it to a small piece of paper towel or blotting material.
- Add approximately 100 µL of PBS, and incubate at room temperature in darkness for 5 min. Remove PBS by absorbing it to a small piece of paper towel or blotting material. Allow the slide to dry slightly (5 to 10 min), but not completely.
- Add slide-mounting medium, and gently place a coverslip on the slide.
- Examine the slide under epifluorescent microscopy. Store unread slides in darkness.

Evaluation of the assay using Echovirus 1 in water samples was conducted in a step-wise fashion, and each experiment involved seeding the virus followed by detection using the ICR method (USEPA, 1996) and the rapid screening assay. In the case of enterovirus concentration using the ICR guidelines, the water is concentrated on 1 MDS filter (ZetaPor Virosorb), eluted in beef extract V (pH 9.5), and further concentrated by flocculation and then re-concentrated by centrifugation. Because the concentrates obtained contain all water components that may affect viruses, it was not necessary to directly seed natural water as it was done for *C. parvum*. Therefore, during the preliminary evaluation experiment, Echovirus 1 was seeded in sterile beef extract V only to investigate potential inhibitory effects of this compound on the rapid screening assay. During the following experiments, different concentrations of Echovirus 1 were seeded in beef extract V containing water concentrates.

For the preliminary experiment, approximately 10,000 PFU of Echovirus 1 were seeded in a 20-mL sample of sterile buffered beef extract V; an unseeded 10-mL sample was used as the experimental negative control (Figure 2.4). Enumeration of the seed solution using a plaque forming assay as described earlier (USEPA, 1996) indicated that only 4,000 PFU were added to the 20-mL aliquot (recovery of 40%). The 20-mL concentrate was split in two 10-mL fractions: one for conventional detection using culture in BGMK cells, and the second for polyethylene glycol (PEG) concentration, RNA extraction and detection using the rapid screening assay.

The PEG concentration involved addition of 2.3 mL of 2.5 M NaCl, and 12.3 mL of sterile 26% PEG 8000 (prepared in DEPC-treated water), to 10 mL of concentrate. The solution was mixed at 4°C for 3 h, and then centrifuged at 7,000xg at 4°C for 30 min. The supernatant was discarded and the pellet containing virus particles resuspended in 600 µL of lysis buffer (S.N.A.P. kit, Invitrogen Corporation, Carlsbad, California). Total RNA was extracted using the S.N.A.P. kit. Serial dilutions of the RNA obtained were performed prior to reverse transcriptase, PCR and the rapid screening assay, leading to Echovirus 1 concentrations of 0.001 – 100 PFU per well. The rapid screening assay utilizing Approach ACM/SL-PRB and colorimetric detection was performed in duplicate wells.

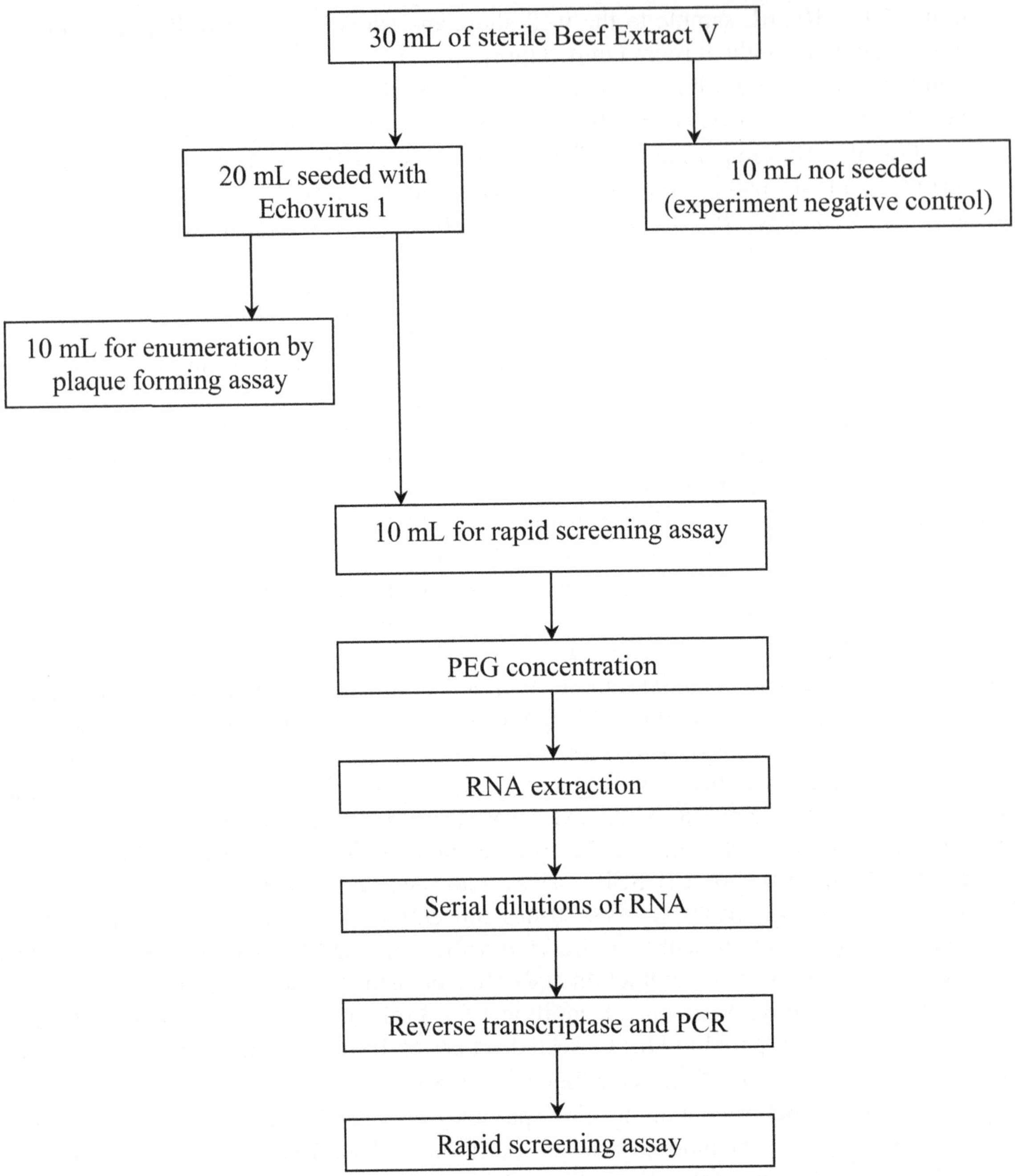


**Figure 2.4. Outline of initial comparison of the rapid screening assay using the amplicon-coated microplate approach (ACM/SL-PRB) and colorimetric detection with the conventional method for enterovirus detection.**

During the following three experiments, six to seven 3-mL aliquots of water concentrates in beef extract V were seeded with different concentrations of Echovirus 1, and one 3-mL aliquot was not seeded (experiment negative control; Figure 2.5). From each 3-mL aliquot, a fraction (0.5 to 1 mL) was used for enumeration by plaque assay, and the remaining fractions (2 to 2.5 mL) were used for PEG concentration, RNA extraction and rapid screening assay. Serial dilutions of the RNA were performed prior to reverse transcriptase and PCR, leading to various Echovirus 1 concentrations as presented in Table 2.7. Approach ACM/SL-PRB and colorimetric detection were used on replicate samples (duplicate RNA samples, and duplicate cDNA samples for each reverse transcriptase reaction). The rapid screening assay was also repeated using the same PCR samples, leading to duplicate assay measurements.

## CUSTOM LIGHT DIAGNOSTICS OLIGODETECT KIT

The OligoDetect kit (Chemicon International, Inc.) is a commercially available hybridization capture assay that detects amplified nucleic acid specific to the target organism of interest. The reported sensitivity is 5 copies of amplified DNA, or 1.5 femtomoles. Through a collaboration with Chemicon, a customized version of the OligoDetect kit was developed for specific detection of *C. parvum*. This customized kit used the same primers (CPHSP2423F and CPHSP2764R) and probe (CPHSP2475) sequences that were used during the development and optimization of the rapid screening assay, as previously developed by Rochelle *et al* (1997b) and described in Table 2.2. The OligoDetect kit is provided with recommendations about the hybridization conditions, although these conditions can be modified to improve sensitivity and/or specificity. Amplification reactions were similar to those used for the rapid screening assay and contained: 1X PCR buffer, 1.5 mM $MgCl_2$, 200 μM each dNTP, 0.30 μM Primer Mix, 0.01 U/μL UNG, 0.025 U/μL AmpliTaq Gold, and 20 μL of nucleic acid template in a final reaction volume of 100 μL. A relatively high concentration of BSA was used (200 μg/μL); such high level may induce non-specific amplification, although verification of the assay specificity indicated that this was not a problem (see Chapter 3). Aliquots of 20 μL DNA sample per 100 μL reaction were used. The PCR activity included 40 cycles of denaturation at 94°C for 1 min, annealing at 60°C for 30 sec, and extension at 72°C for 1 min. To verify amplification efficiency prior to hybridization, amplicons were separated by agarose gel electrophoresis, stained with ethidium bromide, and photographed (Sambrook *et al.*, 1989).

**Table 2.7**
**Concentrations of Echovirus 1 tested during evaluation of Approach ACM/SL-PRB with seeded samples**

| Aliquot number and RNA dilutions | Echovirus 1 concentration in microplate well (PFU) |
|---|---|
| Aliquot #1 with dilutions | 0.018 to 179 |
| Aliquot #2 with dilutions | 0.028 to 28 |
| Aliquot #3 | 19 |
| Aliquot #4 | 7 |
| Aliquot #5 | 3 |
| Aliquots #6 and 7 | <1 |
| Aliquot #8 | 0 (unseeded) |

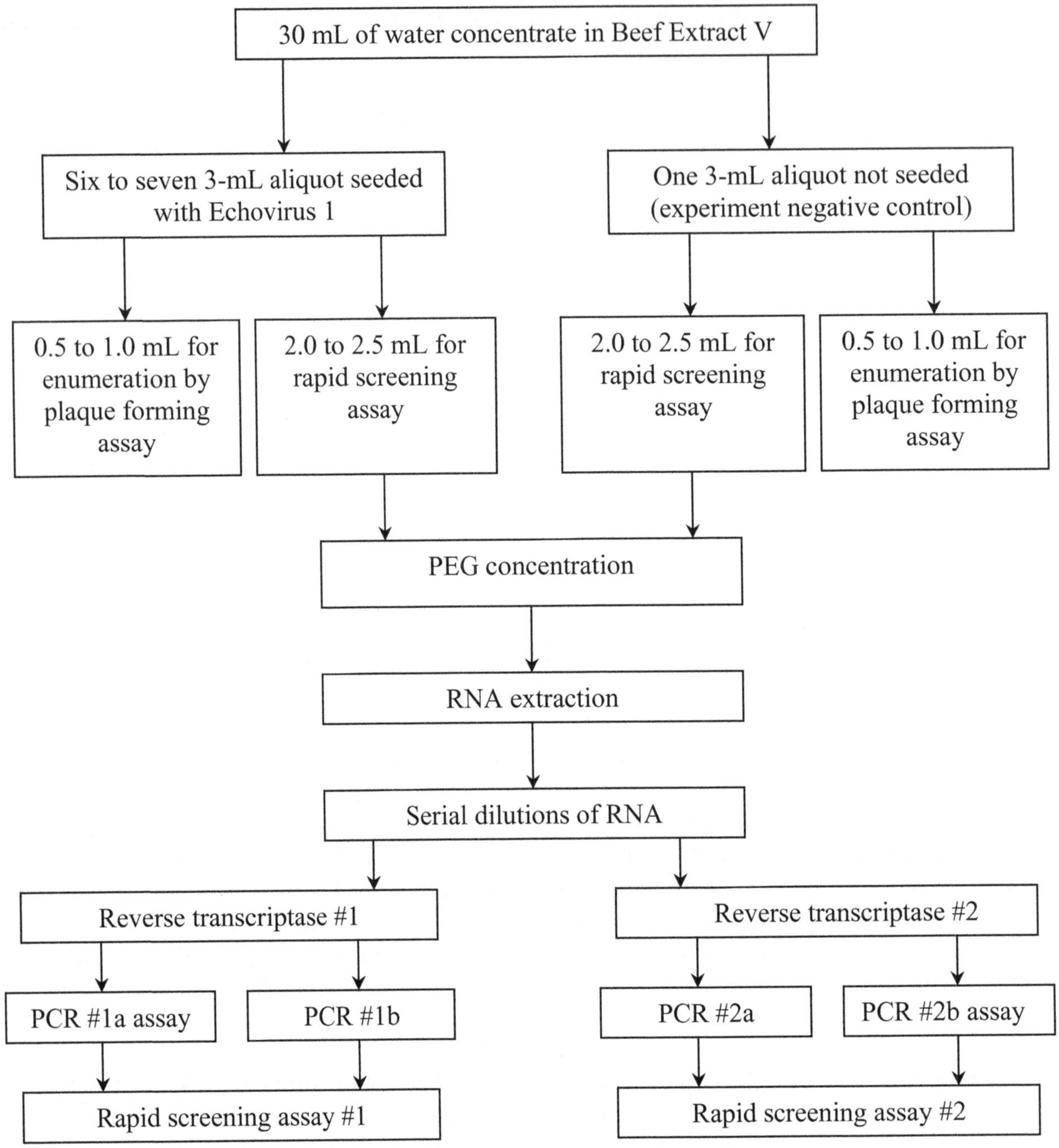


**Figure 2.5. Comparison of the amplicon-coated microplate approach (ACM/SL-PRB) and colorimetric detection with the conventional method for enterovirus detection.**

The microplate hybridization approach used by the OligoDetect kit is similar to a modification of Approach PCM/ML-dUTP developed for this project and is summarized in Figure 2.6. The probe is covalently bound to the 96-well microplate walls so no probe pre-coating period is required. The amplicons are denatured directly in the microplate using an alkaline solution (10 min at room temperature), and the biotin-labeled amplicons hybridize to the attached probe. A hybridization condition of 30 min at 37°C is recommended by the manufacturer. Following hybridization, the microplate is washed in a buffer containing a surfactant and thimerosal.

Colorimetric detection involves incubation at 37°C for 30 min with streptavidin-labeled horse radish peroxidase (HRP), followed by 10 min at 37°C with 3,3',5,5'-tetramethylbenzidine/Enhancer (TMB/E), which forms a blue color at the site of hybridization. The reaction is stopped by the addition of a hydrochloric acid solution, changing the color to yellow. This color is quantified within 5 min using a microplate reader to measure absorbance values at dual wavelengths (450 nm/650 nm), and the color intensity is directly proportional to the amount of amplicon present in the sample. The manufacturer provides the guidelines to evaluate the signal intensity (Table 2.8). The cut-off values were determined for every experiment and were based on replicate analyses of the PCR negative control.

**Table 2.8**
**Guidelines for signal intensity obtained with the OligoDetect Kit**

| Signal | Reading |
|---|---|
| Positive | $A_{450\ nm} \geq x + 3SD$ |
| Negative | $A_{450\ nm} \leq x + 2SD$ |
| Equivocal | $x + 2SD < A_{450\ nm} < x + 3SD$ |

x: Triplicate analyses of PCR negative control
SD: Standard Deviation of triplicate analyses

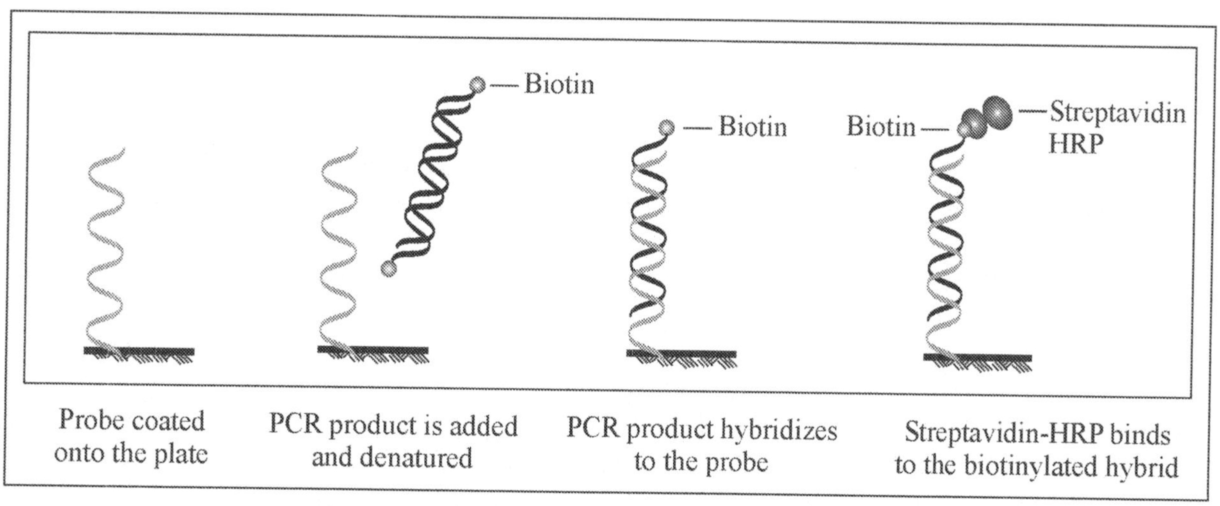


**Figure 2.6. Microplate hybridization and detection approach of a commercially available amplicon detection assay (Custom Light Diagnostics OligoDetect Kit; courtesy of Chemicon International, Inc.).**

## CONTROLS

A variety of controls were conducted during each step of the assay, and are described in the following paragraphs. The positive controls had to show a clear signal and the negative controls had to remain clear in order to determine that a particular experiment was successful.

The structural integrity of the oligonucleotides was tested prior to their utilization. Primer sequences were verified by amplifying previously analyzed DNA extracts of the target organism. The primer specificity was also tested against DNA extracts from other organisms. The presence of biotin on the primers and probes was tested by slot blot on a membrane, followed by detection using streptavidin-AP and CDP-Star chemiluminescent substrate. The presence of DIG on the primers and probes was tested by slot blot with detection using anti-DIG-AP antibody and CDP-Star substrate.

Experiments included both negative and positive controls. At the PCR level, selected experiments included positive controls consisting of an unlabeled amplification (non-labeled dUTP and primers) using the same DNA extracts as that used for the particular experiment. All amplifications included negative controls consisting of sterile molecular biology grade water instead of DNA template. When a nested or double amplification was conducted, both PCRs included negative controls, and the first reaction negative control was also carried through the second amplification. When amplifying echovirus 1, a reverse transcriptase negative control (molecular biology grade water instead of RNA) was incorporated. This negative control was also processed during the PCR, along with a PCR negative control (molecular biology grade water instead of cDNA).

During microplate hybridization, negative controls consisted of sterile molecular biology grade water introduced at the pre-coating step and/or hybridization step (depending on the approach tested), as well as the reverse transcriptase and PCR negative controls from the first and second amplifications if applicable. Additional negative controls were also processed:

1. The pre-coating period was omitted (sterile Milli-Q water was added in the microplate well), and the hybridization, addition of anti-DIG-AP antibody, and colorimetric detection with NBT/BCIP were conducted.
2. The pre-coating period was conducted, the hybridization omitted (sterile Milli-Q water was used instead of probe and hybridization buffer), followed by antibody and detection.
3. The pre-coating and hybridization were conducted, the antibody was omitted (sterile Milli-Q water, or detection buffer with BSA but without antibody in the well), and detection carried through.
4. The pre-coating, hybridization and antibody steps were conducted, but the NBT/BCIP colorimetric substrate omitted.

During Approach ACM/SL-PRB experiments, additional positive controls consisted of biotin-streptavidin-conjugated AP (Jackson ImmunoResearch Laboratories, Inc., West Grove, Pennsylvania; 1.82 and 18.2 μg/mL). The volume of biotin-AP introduced into each microplate well was equal to the sample volume used during the assay (40 μL per well), and was added at one of the following steps:

1. During microplate pre-coating (instead of amplicons).
2. At the hybridization step (the pre-coating was conducted with sterile Milli-Q water).
3. At the antibody step (the pre-coating was performed with sterile Milli-Q water, and no antibody was used).

The biotin-labeled AP was added in the same solution as used for each step (pre-coating buffer, hybridization buffer, or detection buffer with BSA when added at the antibody step). Each slot blot experiment included all negative controls processed during the amplification and microplate assay, as well as sterile water only.

# CHAPTER 3
# RESULTS AND DISCUSSION

A flexible rapid screening assay was developed for the detection of selected pathogens in water. The assay was based on a standard 96-well microplate platform that allowed qualitative results to be recorded by simple visual observation and relative quantification by microplate readers. The assay endpoint was adaptable to both colorimetric and fluorescence-based detection approaches, as outlined in Figures 2.1 and 3.1. Proof of principle and demonstration of feasibility was provided by four selected pathogens (*C. parvum*, *G. duodenalis*, echovirus 1, and *E. coli* O157:H7) but the assay is applicable to any pathogens for which nucleic acid sequence information is available allowing design of specific primers and probes. Similar microplate-based approaches have been used previously for the detection of specific amplification products from poliovirus, hepatitis B virus, human immunodeficiency virus, noroviruses, and *Theileria sergenti* in clinical and cell culture samples (Bon et al., 2004; Erhardt et al., 1996; Scott Eastman et al., 1995; Takeda et al., 1994; Tanaka et al., 1993). The research approach for this project consisted of the following sequential components:

1. Optimization of a prototype method to detect *C. parvum* and evaluation of its adaptability to other microorganisms using echovirus 1 as an example.
2. Evaluating alternative probe hybridization approaches and signal detection techniques using *C. parvum* and echovirus 1 as model microorganisms.
3. Application of the best combination of conditions to two additional microorganisms, *G. duodenalis* and *E. coli* O157:H7.
4. Evaluation of the rapid screening assay with seeded environmental water samples or concentrates.
5. Evaluation of a commercial kit based on similar probe-capture technology.

## OPTIMIZING THE PROTOTYPE ASSAY USING *CRYPTOSPORIDIUM PARVUM*

This section presents the key elements of the optimization performed for the amplification, microplate hybridization and colorimetric detection steps of the prototype assay. *C. parvum* was selected as a model microorganism for assay development because of its relevance to the water industry, as well as the researchers' past experience with this protozoan. The prototype assay involved probe-coated microplates (PCM) and multiple-labeled amplicons (Figures 2.1). The target DNA was amplified by PCR using *C. parvum*-specific primers (Table 2.2) with direct incorporation of DIG-labeled dUTP into the amplicons. The amplicon was transferred into streptavidin-coated 96-well microplates that had been prebound with a 5'-biotin labeled *C. parvum*-specific DNA probe. After hybridization directly in the microplate, unattached amplicons were removed by washing, and the amplicon-probe hybrids were detected with anti-DIG-AP antibody and a colorimetric substrate solution of NBT/BCIP. The anti-DIG antibody binds to the DIG-labeled amplicons and the AP acts on the substrate to produce a purple precipitate in the well at the site of probe hybridization (Figure 3.2). Assay results could be detected visually using qualitative judgment to score signal intensity or by measuring absorbance at 550 nm using a microplate reader to provide relative quantitation. A detailed laboratory protocol for the prototype assay is presented in Appendix A.

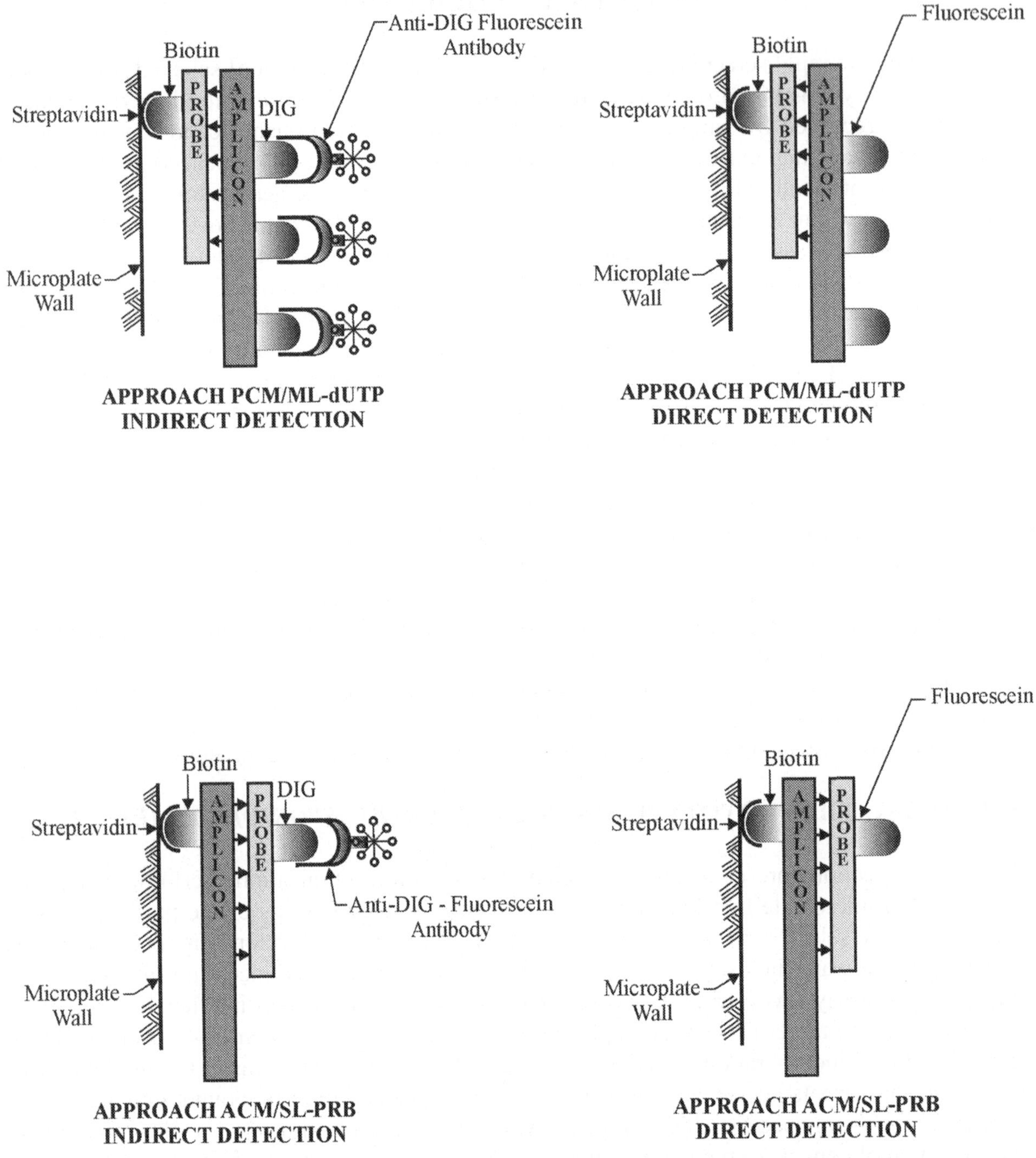


**Figure 3.1. PCR and microplate hybridization approaches for fluorescence-based detection.**

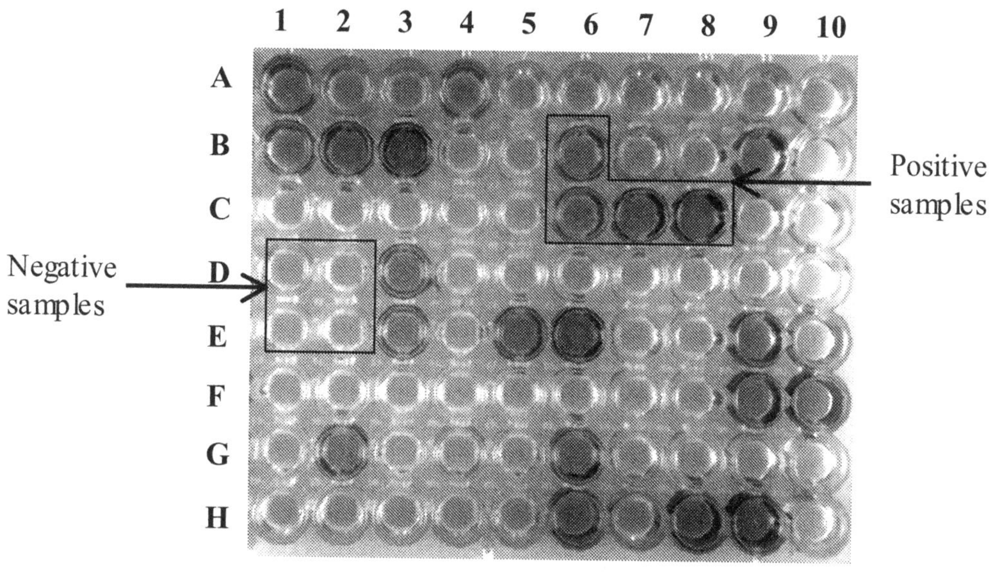


**Figure 3.2. Example of rapid screening assay endpoint with colorimetric detection in 96-well microplate (only 80 wells are displayed in the figure).**

### Optimization of Amplification Conditions

Because the detection of the labeled amplicons depends on both the amount of amplicons produced and the effectiveness of label incorporation during PCR, the amplification conditions were optimized as presented in Table 3.1. Amplification conditions were selected based on optimum sensitivity and specificity after varying PCR parameters individually, and analyzing PCR amplicons by gel electrophoresis and amplicon-probe hybrids in microplates.

In the PCM approach a DIG label was incorporated into the amplicons using DIG-11-dUTP during PCR. Generally, 200 µM each dNTP was used per PCR amplification. However, dUTP is slightly more difficult to incorporate into the amplicons than the other dNTPs and this tendency is emphasized by the presence of DIG on the DIG-11-dUTP molecule. Thus, the overall concentration of dUTP was increased by reducing the concentrations of dATP, dCTP and dGTP (100 µM each dATP, dCTP and dGTP, with 200 µM dUTP). Results indicated stronger signal intensity with these conditions (Table 3.2).

The ratio of DIG-11-dUTP:dUTP was optimized as more DIG-11-dUTP resulted in a lower amount of PCR products by gel electrophoresis, but stronger signal intensity by hybridization in a microplate. Results indicated that the optimum ratio of DIG-11-dUTP:dUTP was 1:4 (40 µM DIG-11-dUTP and 160 µM non-labeled dUTP; Table 3.3). This evaluation was conducted jointly with the optimization of the $MgCl_2$ concentration to confirm that the optimum $MgCl_2$ concentration necessary when PCR included DIG-11-dUTP was similar to that stated in the literature.

**Table 3.1**
**PCR conditions evaluated for *C. parvum* detection using the probe-coated microplate (PCM) approach**

| PCR conditions | Range tested | Optimum or selected conditions |
|---|---|---|
| Ratio of dUTP to other dNTPs | 1:1, 2:1 | 2:1 (200 μM dUTP, 100 μM each dATP, dCTP, dGTP) |
| DIG-11-dUTP:dUTP ratio | 1:4 – 1:20 | 1:4 |
| $MgCl_2$ concentration | 1.5 – 3.5 mM | 1.5 mM |
| Primer concentration | 0.25 and 0.50 μM | 0.50 μM each primer |
| Annealing temperature | 52 and 55°C | 55°C |
| Amplification | Single and double | Single |

**Table 3.2**
**Summary of signal intensity obtained with different dNTP concentrations**

| Number of oocysts in PCR | 100 μM dATP, dCTP, dGTP, with 200 μM dUTP | 200 μM each dNTP |
|---|---|---|
| $5x10^5$ | ++ | +++ |
| $5x10^4$ | ++ | ++ |
| $5x10^3$ | + | + |
| $5x10^2$ | + | - |
| $5x10^1$ | - | - |
| PCR negative control | - | - |

Note: -, no signal; +/-, weak signal; + to ++++, pale to darkest signal.

**Table 3.3**
**Summary of signal intensity obtained with different DIG-11-dUTP:dUTP ratios and $MgCl_2$ concentrations**

| | Total Concentration of $MgCl_2$: | | |
|---|---|---|---|
| Ratio DIG-11-dUTP:dUTP | 1.5 mM | 2.5 mM | 3.5 mM |
| 1:4 | +++ | ++ | + |
| 1:6 | ++ | na | na |
| 1:8 | ++ | + | + |
| 1:10 | + | na | na |
| 1:12 | - | - | - |
| 1:15 | - | na | na |
| 1:20 | - | na | na |

Note: -, no signal; +/-, weak signal; + to ++++, pale to darkest signal.
na: Not analyzed.

## Optimization of Microplate Hybridization Conditions

Following amplification, the presence of *C. parvum* was confirmed by hybridization in microplates. A previously published probe (CPHSP2475; Rochelle et al., 1997b) was modified to work as a capture probe by the addition of a 5'-biotin label. The biotin-labeled capture probe bound to the streptavidin coating of the microplate. The amplicons were heat denatured (95°C for 5 min followed by 99°C for 1 min), and hybridized to the probe directly in the microplate.

Method development was performed to improve the assay sensitivity while preserving its specificity. It included testing of different incubation periods, concentrations of hybridization reagents and probe washing solutions (Table 3.4), and evaluation of the resulting signal intensity in microplate wells. The concentrations of the different components of the buffer solution (blocking reagent, SSC, and SDS) were evaluated during both the pre-coating and hybridization steps. A microplate pre-coating period was introduced to allow the biotin-labeled probe to bind to the streptavidin coating of the microplate. It was anticipated that this pre-coating step would increase the amount of probe attached to the microplate walls. During this step, the hybridization buffer (SSC, blocking reagent, SDS) containing the probe was left in contact with the microplate wells prior to the introduction of the amplicons (Table 3.5). This evaluation was performed in conjunction with the optimization of the hybridization period.

Stronger signal intensity was obtained by increasing the amount of amplicon (volume of PCR) transferred to the microplate wells. The aim of the method was to amplify and hybridize the DNA of an unknown sample to detect the presence of a specific pathogen. Amplification in 50-μL reactions is currently widely used and provides reliable and reproducible results at a reasonable cost. Therefore, transferring 40 μL of the amplified sample leaves the remaining volume available for gel electrophoresis or further analyses if necessary.

Signal intensity was stronger with increasing salt concentration (Table 3.6). However, the higher salt concentrations tested (SSC levels of 8X and 10X) led to non-specific background. Thus, an SSC concentration of 6X was selected. Increasing probe concentration up to 100 pM did not improve signal intensity. Stronger signal intensity was obtained by increasing the hybridization period, although maximum signal was obtained after a 2 to 3-hour hybridization period (Table 3.5). Additional hybridization time did not significantly improve the signal intensity.

**Table 3.4**
**Hybridization conditions evaluated for *C. parvum* detection using the PCM approach**

| Hybridization conditions | Range tested | Optimum or selected conditions |
|---|---|---|
| Pre-coating period | 0 – 4.25 h | 30 min |
| Probe concentration | 20 – 100 pM | 40 pM |
| Sample (amplicon) volume | 10 – 50 μL | 40 μL |
| SSC concentration | 0 – 10X | 6X |
| Blocking reagent concentration | 1 – 4% | 1% (v/v) |
| SDS concentration | 0 – 0.02% | 0.02% (w/v) |
| Hybridization period (59°C) | 0.5 – 21 h | 2 – 3 h |
| Post-hybridization washing temperature | 57 and 60°C | 57°C |
| Post-hybridization washing solution | (1) 5X SSC<br>(2) 5X SSC, 20 mM Tris-HCl (pH 7.5)<br>(3) 6X SSC, 0.02% SDS<br>(4) 5X SSC, 0.1% (w/v) SDS, then 5X SSC, 0.1% (w/v) Triton-X-100, then 5X SSC | 5X SSC, 20 mM Tris-HCl (pH 7.5) |

Note: 1X SSC contains 0.15 M NaCl and 0.015 M sodium citrate.

**Table 3.5**
**Summary of signal intensity obtained with different pre-coating and hybridization periods**

| Time (h) | Pre-coating period | Hybridization period |
|---|---|---|
| 0.5 | +++ | + |
| 1 | na | ++ |
| 1.5 | ++ | +++ |
| 2 | ++ | +++ |
| 3 | na | +++ |
| 4 | + | +++ |
| 18-22 | na | +++ |

Note: -, no signal; +/-, weak signal; + to ++++, pale to darkest signal.
na: Not analyzed.

**Table 3.6**
**Summary of signal intensity obtained with different pre-coating and hybridization buffer composition**

| SSC | | SDS | | Blocking reagent | |
|---|---|---|---|---|---|
| 0X | - | 0% | + | 1% | +++ |
| 2X | + | 0.01% | ++ | 2% | ++ |
| 4X | ++ | 0.02% | +++ | 4% | + |
| 6X | +++ | | | | |

Note: -, no signal; +/-, weak signal; + to ++++, pale to darkest signal.
1X SSC contains 0.15 M NaCl and 0.015 M sodium citrate.

**Optimization of End-Point Colorimetric Detection**

Detection of DIG-labeled amplicon-probe hybrids was conducted in a conjugate solution consisting of 1% (w/v) BSA in detection buffer containing anti-DIG-AP antibody. The conjugate solution was incubated at room temperature for 30 min with gentle agitation. To determine if signal intensity could be further improved, two antibody concentrations were tested, 0.75 and 1.5 U/mL. Results indicated non-specific background at the higher concentration. Therefore, an antibody concentration of 0.75 U/mL (0.1%, v/v) was selected. The visual colorimetric detection was conducted with a substrate solution of 2% (v/v) NBT/BCIP in detection buffer. Samples positive for *C. parvum* developed a visible purple color within 15 min at room temperature, whereas negative samples remained clear (Figure 3.2). To increase signal intensity, different incubation periods in NBT/BCIP substrate were tested (15 – 150 min) and the optimum duration was established at 120 min at room temperature. Color intensity increased for approximately 2 h, but longer incubation periods resulted in non-specific background.

**Assay Sensitivity**

Using the optimized amplification and microplate hybridization conditions, the best sensitivity achieved was 5 oocysts (evaluated by serial dilutions of hemocytometer enumerated stock oocyst preparations).

**Assay Specificity**

The specificity of the *C. parvum*-specific primer/probe combination targeting the hsp70 gene was demonstrated in earlier studies using membrane hybridization (Rochelle et al., 1997b; Rochelle et al., 2002). The primers amplified the intended target from all isolates of *C. parvum* tested (Iowa, Maine, Moredun, Glasgow, TAMU) as well as isolates of *Cryptosporidium hominis* (previously referred to as genotype 1 *C. parvum*). The primer/probe combination was negative with all of the following organisms, representing protozoa, bacteria, algae, and mammalian cell lines: *Cryptosporidium muris*, *Cryptosporidium baileyi*, *Giardia duodenalis*, *Giardia muris*, *Hexamita* sp., *Entamoeba histolytica*, *Entamoeba coli*, *Babesia microti*, *Endolimax nana*, *Blastocystis hominis*, *Trypanosoma theileri*, *Cyclospora* sp., *Toxoplasma gondii*, *Dientamoeba fragilis*, *Eimeria maxima*, *Acanthamoeba castellani*, *Pneumocystis carinii*, and *Encephalitozoon* sp., *Escherichia coli*, *Pseudomonas aeruginosa*, *Aeromonas hydrophila*, *Enterobacter aerogenes*, *Klebsiella pneumoniae*, *Bacillus subtilis*, *Synechococcus* sp., *Synechocystis* sp., *Chlorella vulgaris*, *Selenastrum capricornutum*, *Scenedesmus obliquus*, *Saccharomyces cerevisiae*, and mammalian cell lines (Caco-2, HCT-8, MDBK, and Vero). These experiments were conducted under the same PCR amplification conditions ($MgCl_2$ concentration and annealing temperature of 55°C), hybridization and post-hybridization washing temperature conditions, as used during the rapid screening assay. Results of Rochelle et al (1997b), as well as from other in-house experiments, demonstrate that the primers and probe are specific for several isolates and genotypes of *C. parvum*.

The specificity of the rapid screening assay developed for this project was further verified under the selected conditions for Approach PCM/ML-dUTP and colorimetric detection. The microorganisms tested included (refer to Table 2.1 for sources of microorganisms): *Aeromonas hydrophila*, *E. coli* O157:H7, non-toxigenic *E. coli*, *Klebsiella pneumoniae*, *Salmonella typhimurium*, *Nitrosomonas europea*, *Comamonas testosteroni*, and *Saccharomyces cerevisiae*. Results confirmed the specificity of the *C. parvum* primers/probe combination when used in the PCM approach of the rapid screening assay.

### Additional Issues Evaluated with the Prototype Assay for Detection of *Cryptosporidium parvum*

Results from multiple experiments conducted with the PCM approach on *C. parvum* indicated that the most critical parameters for optimum performance were the ratio of DIG-11-dUTP:dUTP, the sample volume, the SSC concentration, and sufficient contact time with the NBT/BCIP colorimetric substrate. Several parameters such as the polymerase enzyme, SDS and blocking reagent concentrations, and hybridization period, had a negligible impact on the assay performance. Other parameters or conditions were also evaluated and some of them are briefly presented below, but results indicated that they did not improve the assay sensitivity nor specificity.

#### *Optimization of Amplification Conditions*

A lower annealing temperature was investigated (52°C compared to 55°C) to improve the assay sensitivity, but did not result in a significant improvement. Because a lower annealing temperature also decreases the specificity, the annealing temperature was fixed at 55°C.

Early experiments indicated that two successive rounds of amplification (double PCR or semi-nested PCR) may be necessary to increase the assay sensitivity. Nonetheless, once the assay was optimized and High-Bind streptavidin-coated microplates were used, acceptable sensitivity was obtained with single PCR amplification.

#### *Optimization of Microplate Post-Hybridization Conditions*

In order to decrease background signal, various post-hybridization washing solutions were evaluated (Table 3.4) based on a thorough review of commercially available washing buffers, literature, and molecular biology handbooks such as Hames (1986) and Howard (1993). Results indicated that SDS-based washing solutions led to deposition of colored precipitate in the negative controls. According to Howard (1993), SDS causes chemical deposition of the NBT/BCIP dye and should therefore be avoided prior to substrate addition. This observation justified our choice of buffer (2) (5X SSC and 20 mM Tris-HCl [pH 7.5]), which was also the buffer used during all previous experiments.

No significant difference was observed between the two post-hybridization washing temperatures (57 and 60°C). A temperature of 57°C in combination with high SSC concentration was adopted to improve sensitivity. Also, 57°C is the temperature generally used during the post-hybridization washing step of conventional membrane hybridization with this probe (Rochelle et al., 1997b).

*Primer and Probe Haptens*

The assay sensitivity may depend on the effectiveness of the probe to successfully capture the amplicon, which is linked to the ability of the probe-amplicon hybrids to hybridize together in the presence of a hapten, as well as once attached to the streptavidin coating of the microplate wells. To determine if stearic hindrance was reducing the assay sensitivity, probes were manufactured with a 15-carbon chain link between the biotin molecule and the first probe nucleotide (BIOTEG). Using Approach PCM/ML-dUTP and its modified version (Approach PCM/SL-PRM) for the colorimetric detection of *C. parvum*, the 15-carbon probe was compared to the two-carbon chain link probe previously used. Results showed no significant difference between both probes, which indicated that stearic hindrance is not an issue in the assay, or occur at the same level for both types of probe.

To further evaluate the potential for stearic hindrance, a probe containing a chain of six adenine bases between the biotin label and the probe sequence (5'-poly-A probe) was compared to the 15-carbon chain link biotin-labeled probe using Approach PCM/ML-dUTP and the colorimetric detection for *C. parvum*. Results showed that the 5'-poly-A probe provided the same method sensitivity as the 15-carbon biotin-labeled probe (Table 3.7), therefore suggesting that stearic hindrance does not occur, or occur at the same level with both probes.

*Modification of the Prototype Assay*

PCR amplification with DIG-11-dUTP was significantly less efficient than non-labeled amplification, and therefore may reduce the method sensitivity. This was observed by the absence of a visible band on the gel electrophoresis when PCR was performed with DIG-11-dUTP. Also, DIG-11-dUTP is an expensive reagent. Thus, a modification of Approach PCM/ML-dUTP (named Approach PCM/SL-PRM; Figure 2.1) was proposed to rely on the efficiency of conventional PCR using unlabeled dNTPs and placing a single DIG label (SL) at the 5'end of the primers (PRM). The resulting DIG-labeled amplicons were transferred into 96-well, streptavidin-coated microplates with a pre-attached, biotin-labeled probe. After hybridization in the microplate, unattached amplicons were removed by washing, and the amplicon-probe hybrids were detected with the colorimetric method.

Selected amplification and hybridization conditions were tested to improve assay sensitivity while maintaining specificity (Table 3.8), and the results were analyzed using gel electrophoresis of the labeled amplicons and signal intensity in the microplates following hybridization.

**Table 3.7**
**Summary of signal intensity obtained with different types of probe-hapten linkers**

| Number of oocysts in PCR | 5'-poly-A linker | 15-carbon linker |
|---|---|---|
| 500 | ++ | ++ |
| 50 | ++ | ++ |
| 5 | +/- | +/- |
| <1 | - | - |
| Negative control | - | - |

Note: -, no signal; +/-, weak signal; + to ++++, pale to darkest signal.

**Table 3.8**
**PCR and hybridization conditions evaluated for *C. parvum* detection using Approach PCM/SL-PRM**

| Conditions | Range tested | Optimum or selected conditions |
|---|---|---|
| **PCR conditions:** | | |
| Primer sets | CPHSP2423F/CPHSP2764R+DIG<br>CPHSP2423F+DIG/CPHSP2764R<br>CPHSP2423F+DIG/CPHSP2764R+DIG | CPHSP2423F/<br>CPHSP2764R+DIG |
| Amplification | Single and double | Single |
| **Hybridization conditions:** | | |
| SSC concentration | 4X and 6X | 6X |
| SDS concentration | 0.01 and 0.02% | 0.02% |
| Blocking reagent concentration | 1 and 3% | 1% |

Theoretical DNA sequence analysis indicates that the primer set CPHSP2423F (positive polarity) and CPHSP2764R+DIG (negative polarity), or CPHSP2423F+DIG and CPHSP2764R+DIG would provide optimum results as the hybridization probe (CPHSP2475; positive polarity) anneals to the negative amplicon strand generated by the negative CPHSP2764R primer only, while the positive strand generated by the positive CPHSP2423F primer is lost during denaturation and subsequent washing. Nonetheless, all three possible primer combinations were empirically evaluated for specificity comparison purposes. As expected, optimum results were obtained with the expected primer sets (no signal was obtained with primer set CPHSP2423F+DIG and CPHSP2764R). All experiments were conducted with 40 pM of biotin-probe. The amplification and hybridization optimum conditions were similar to those determined for the Approach PCM/ML-dUTP. The signal was however much weaker with the Approach PCM/SL-PRM than with the original Approach PCM/ML-dUTP. It appears that the smaller amount of DIG molecules per amplicon obtained with the Approach PCM/SL-PRM was not sufficient to allow adequate detection of the amplicon/probe hybrids.

## ALTERNATIVE AMPLIFICATION AND HYBRIDIZATION APPROACHES FOR *CRYPTOSPORIDIUM PARVUM*

Alternative amplification and microplate hybridization approaches were evaluated as a means to possibly increase the robustness of the assay, improve sensitivity, and decrease the cost and duration of analysis. The alternative amplification and hybridization approaches evaluated are presented on Figures 2.1 and 3.1. In Approach ACM/SL-PRB, biotin is incorporated in the amplicons during amplification by the use of biotin-labeled primers. Thus the amplicons were bound to the microplate streptavidin coating. The probe contains a hapten to which a labeled antibody bound.

During most experiments, DIG-labeled probes were used with the same colorimetric detection method as for Approach PCM/ML-dUTP (anti-DIG-AP antibody and NBT/BCIP substrate). A detailed laboratory protocol is presented in the Appendix B.

## Optimization of Amplification Conditions

Several conditions were tested to increase the amount of amplicon generated during PCR amplification with biotin-labeled primers (Table 3.9). Results were analyzed using gel electrophoresis of the PCR amplicons and signal intensity in the microplate after hybridization.

For the PCR conditions not affected by the amplicon labeling process, the conditions used for Approach PCM/ML-dUTP were also employed for Approach ACM/SL-PRB: 1.5 mM $MgCl_2$, 200 µM each dNTPs with dUTP substituted for dTTP so that UNG (0.01 U/µL) could be used, and 0.01 µg/µL BSA.

Although the three possible primer sets were tested for comparison purposes, theoretical DNA sequence analysis indicates that only the negative amplicon strands generated with the reverse primer (CPHSP2764R, negative polarity) can hybridize with the probe (positive polarity). Thus, most experiments were conducted with primer CPHSP2423F and CPHSP2764R+biotin (Table 3.9) to prevent the unnecessary attachment to the microplate of the positive-polarity amplicon strand which cannot hybridize with the probe. The use of only one labeled primer also reduces the cost of the assay. Nonetheless, subsequent assay optimization demonstrated that sufficient sensitivity was achieved with both forward and reverse primers labeled with biotin.

Assay specificity was evaluated using the organisms described in Chapter 2 and as expected was the same for the ACM approach as was demonstrated for the PCM approach.

Lower primer concentrations were examined (50, 100 and 250 nM, compared with 500 nM) in attempt to decrease the amount of unincorporated biotin-labeled primers potentially attaching to the microplate. No observable differences were obtained between the primer concentrations tested, although the signal intensity was weaker for the lower *C. parvum* oocyst concentrations analyzed with the lower primer concentrations. Therefore, a concentration of 250 nM was adopted as this is the concentration generally used in common PCR amplification.

Different polymerase enzymes were evaluated: AmpliTaq and AmpliTaq Gold (Perkin-Elmer Corp.); Platinum Taq (Gibco BRL); Tth and Expand High Fidelity (Roche Molecular Biochemicals). Results indicated that for the amplification of *C. parvum* DNA using biotin-labeled primers and hybridization with DIG-labeled probe in a microplate, AmpliTaq and AmpliTaq Gold seemed to provide stronger signal than Platinum, Tth or Expand enzymes. Because AmpliTaq Gold has higher specificity than AmpliTaq, it was thus selected for Approach ACM/SL-PRB.

It has been demonstrated that the dUTP/UNG system is efficient to prevent contamination from previous amplification (Longo et al, 1990; Schmidt et al, 1996). However, in-house experiments have shown that the system can interfere with cloning, restriction enzyme digestion, and destroy amplicons during storage even at low temperature (4°C). Thus experiments were conducted to compare amplification using dTTP vs. the dUTP/UNG system. Results obtained indicated that although dTTP provided stronger signal than the dUTP/UNG system (Figure 3.3), the dUTP/UNG system provided sufficient sensitivity and was preferred as it prevents contamination from previous amplification.

**Table 3.9**
**PCR conditions evaluated for the detection of *C. parvum* using Approach ACM/SL-PRB**

| PCR parameter | Range tested | Optimum or selected conditions |
|---|---|---|
| Primer sets | 1. CPHSP2423F+biotin/CPHSP2764R<br>2. CPHSP2423F/CPHSP2764R+biotin<br>3. CPHSP2423F+biotin/CPHSP2764R+biotin | 2 or 3 |
| Primer concentration | 50 – 500 nM | 250 nM |
| Polymerase enzyme | AmpliTaq, AmpliTaq Gold, Platinum Taq, Tth, and Expand Taq | AmpliTaq Gold |
| dNTPs | dUTP and UNG, or dTTP | dUTP and UNG |
| Amplification | Single or double | Single |

**Optimization of Microplate Hybridization Conditions**

Following amplification, amplicons were denatured at 95°C for 5 min and then at 99°C for 1 min. Denatured amplicons were transferred to microplate wells for attachment to the streptavidin coating during a pre-coating period. Hybridization was conducted directly in 96-well microplates using a 5'-DIG labeled CPHSP2475 probe. Several hybridization parameters were evaluated as presented in Table 3.10. The optimum salt concentration was determined at 5 to 6X SSC, with only a very subtle difference noted between these two concentrations. Some negative controls showed non-specific signal when higher salt concentrations (above 6X) were used during the microplate hybridization. Different probe concentrations were also examined and optimum results were obtained with the higher concentrations of 10 nM.

Incorporation of additional label into the probe with an oligonucleotide tailing kit was also investigated as a means to improve the assay sensitivity, as it was suspected that the low amount of label on the probe (only one molecule per probe) might be a limiting step. This kit adds multiple labels (DIG or fluorescein) at the 3'-end of the probe. Much stronger signal was obtained with probes treated with the oligonucleotide tailing kit than with the standard probes (Table 3.11). Supplemental experiments involving probe binding on a slot blot with Nylon membrane and chemiluminescence detection, demonstrated that the oligonucleotide tailing kit was beneficial and added a significant amount of label to the probes (Figure 3.4). However, for the purposes of this study, the single 5'-labeled probes were used for the majority of assay optimization since they provided the same sensitivity as the multiple 3'-tailed probes, although the signal was not as strong (Figure 3.4). Also, probe tailing adds an additional series of manipulations and time to the assay. Nevertheless, probe tailing provides an extra level of sensitivity that may be necessary with very low numbers of target organisms or to compensate for sample inhibition.

A probe fluorescein hapten was compared to the DIG hapten. Results indicated that with Approach ACM/SL-PRB and *C. parvum*, the DIG-label probe provided more sensitive and reproducible results than the fluorescein-label probe (Table 3.11).

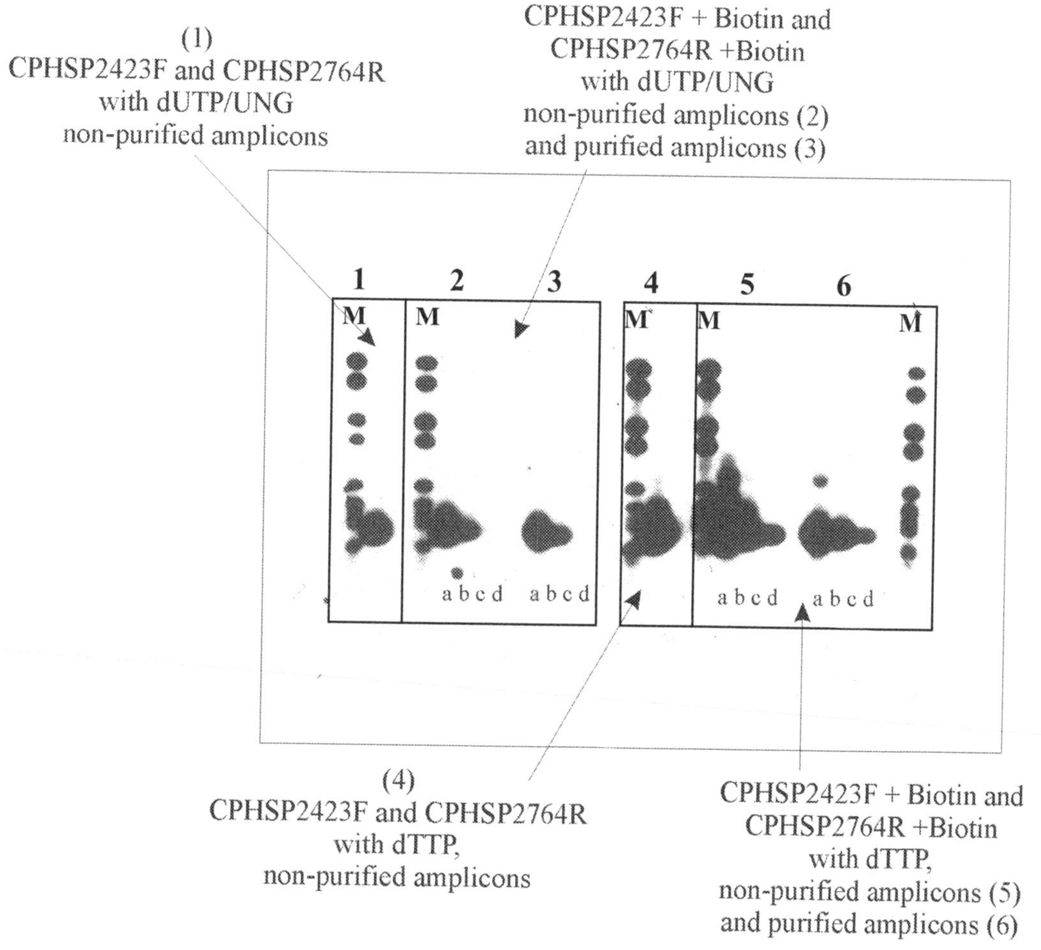


**Figure 3.3. Southern transfer of purified and non-purified PCR amplicons generated with various primer combinations and dNTP strategies. Detection was performed using streptavidin-alkaline phosphatase conjugate and CDP-Star chemiluminescent substrate. In panels 1 and 4 the left well is the positive control (approximately 3,100 oocysts in PCR) and the right well is the PCR negative control. In panels 2, 3, 5, and 6, amplification reactions contained DNA from (a) 6,100 oocysts, (b) 610 oocysts, (c) 61 oocysts, and (d) negative control. M, Labeled molecular size standard.**

**Table 3.10**
**Hybridization conditions evaluated for *C. parvum* detection using Approach ACM/SL-PRB**

| Hybridization conditions | Range tested | Optimum or selected conditions |
|---|---|---|
| Amplicon denaturation and pre-coating scenarios | (1) Denaturation at 95°C for 5 min, then 99°C for 1 min (in PCR tubes); Pre-coating for 0 – 2 h at room temperature or 4°C (in microplate).<br>(2) Pre-coating for 1 h at room temperature (in microplate); Denaturation in 25 or 50% formamide for 1 h at 55°C (in microplate).<br>(3) Pre-coating for 1 h at room temperature (in microplate); Denaturation for 5 – 30 min at room temperature in microplate in 0.015 M NaOH (pH 12), or 0.15 M NaOH (pH 13), or 0.4 M NaOH and 1.5 M NaCl (pH 13). | Denaturation for 95°C for 5 min, then 99°C for 1 min; Pre-coating for 1 h at 4°C |
| Amplicon pre-coating buffer | (1) Sterile Milli-Q water<br>(2) Tris buffer (pH 9.5)<br>(3) 1.0 M Tris-HCl (pH 7.5)<br>(4) SSC, SDS, and blocking reagent | 5X SSC, 0.02% SDS, and 1% blocking reagent |
| Probe concentration | 0.02 – 10 nM | 10 nM |
| Blocking reagent concentration | 0 – 4% | 1% |
| SSC concentration | 2 – 10X | 6X |
| SDS concentration | 0 – 0.04% | 0.02% |
| Hybridization period (59°C) | 2 – 20 | 2 – 3 h |

**Table 3.11**
**Summary of signal intensity obtained with standard labeled probes and probes treated with the Oligonucleotide Tailing kit (DIG and fluorescein labels)**

| | Fluorescein-labeled probes | | DIG-labeled probes | |
|---|---|---|---|---|
| Number of Oocysts in PCR | Standard | Oligonucleotide Tailing kit | Standard | Oligonucleotide Tailing kit |
| 10,000 | +/- | + | + | +++ |
| 1,000 | +/- | + | + | ++ |
| 100 | - | - | - | - |
| Negative control | - | - | - | - |

Note: -, no signal; +/-, weak signal; + to ++++, pale to darkest signal.
Results obtained with 100 pM probe concentration.

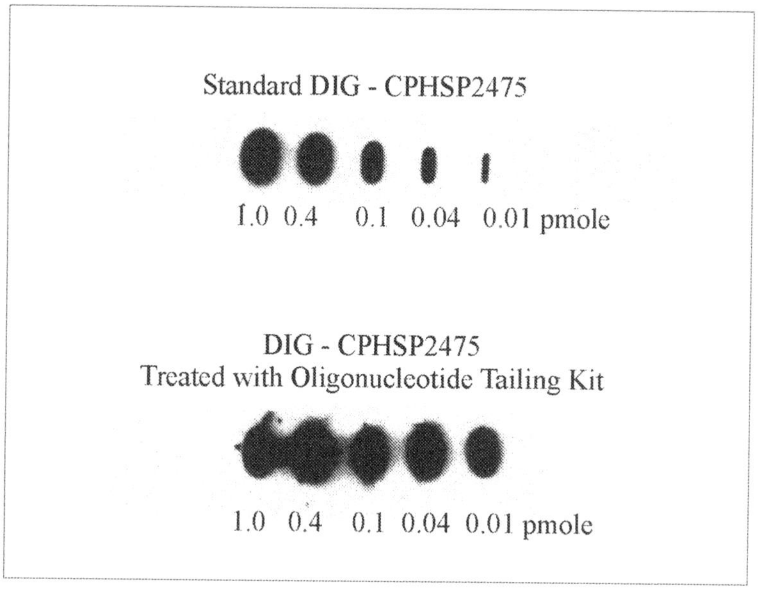


**Figure 3.4. Labeling efficiency of *C. parvum*-specific probes containing a single DIG label (standard) compared to multiple DIG labels added with an oligonucleotide tailing kit. Probes were detected using anti-digoxigenin-alkaline phosphatase and CDP Star chemiluminescent substrate without hybridization.**

***PNA Probes***

PNA is a synthetic DNA analogue that binds to DNA molecules the same way that DNA-DNA hybrids form. PNA has many of the properties of DNA without some of its disadvantages (Table 3.12). PNA-DNA hybrids exhibit a high thermal stability because of the PNA probe's electrical neutrality. As opposed to DNA-DNA hybrids, the $T_m$ of the PNA-DNA hybrids is almost independent of salt concentration. Consequently, PNA-DNA hybrids form with greater specificity and affinity than DNA-DNA hybrids. The $T_m$ is generally higher, approximately 1°C per bp, leading to $T_m$ in the order of 70°C. PNA probes are usually shorter than DNA probes, 15 bases or less, allowing detection of microorganisms that exhibit high sequence variability and providing greater specificity and affinity. Other potential advantages of the PNA probes are rapid hybridization times (5 – 30 min), resistance to enzymatic degradation, preservation of cellular integrity in whole-cell hybridization, identification of point mutation, ability to run PNA-DNA hybrids on electrophoresis gels, and capability of forming a very stable triplex PNA-DNA-PNA hybrids. (Egholm et al., 1993).

**Table 3.12**
**Comparison between DNA and PNA probes**

| PNA probes | DNA probes |
|---|---|
| Neutral charge: No repulsion by DNA targets during hybridization | Negative charge: Repulsed by negatively charged DNA targets during hybridization |
| Hybridization less dependent on salt concentration | Hybridization highly dependent on salt concentration |
| Short probes (10-18 bases) | Longer probes (16-34 bases) |
| Rapid hybridization (5 - 30 min) | Slow hybridization (4 - 18 hours) |
| More specific hybridization: Single base mismatch not tolerated | Non-specific hybridization possible: Multiple base mismatches possible |

The PNA probe format was examined using our model microorganism, *C. parvum*, and the most suitable and practical approach, Approach ACM/SL-PRB. The probe was designed with a fluorescein label to allow both indirect colorimetric detection (with an anti-fluorescein-AP antibody and NBT/BCIP substrate) and direct fluorescent detection. Prior to their utilization in microplates, the fluorescein-labeled PNA probe was tested by slot blot hybridization on a Nylon membrane under different conditions. Considering that PNA probe hybridization is independent of salt concentration, it is possible to simply use a Tris-EDTA solution as hybridization buffer. However, concerns were raised that such simple buffer might not be suitable for hybridization in a microplate. Therefore, two hybridization buffer compositions (1x Tris-EDTA buffer vs. a solution of 1x SSC, 1% blocking reagent and 0.02% SDS) and two hybridization temperatures (room temperature and 37°C) were evaluated using slot blot hybridization. All conditions evaluated included a 30-min hybridization, and were compared to a standard overnight hybridization using DNA probe. Results indicated that the PNA probe is much more sensitive (>1 log difference based on hemocytometer counts) than its DNA counterpart (Figure 3.5). However, the Tris-EDTA buffer did not provide as much specificity as the SSC-Block-SDS solution, as a pale signal was observed in the hybridization negative control. Dot blot hybridization at higher temperature (37°C) improved the sensitivity of ~1 log compared to room-temperature hybridization (Figure 3.5).

To investigate the possibility that the biotin label on the amplicons may inhibit hybridization with the PNA probe, an additional test was conducted during which biotin-labeled amplicons were Southern transferred to a Nylon membrane and hybridized to the PNA probe. Results indicated that the biotin-labeled amplicons were capable of hybridization to the PNA probe (Figure 3.6).

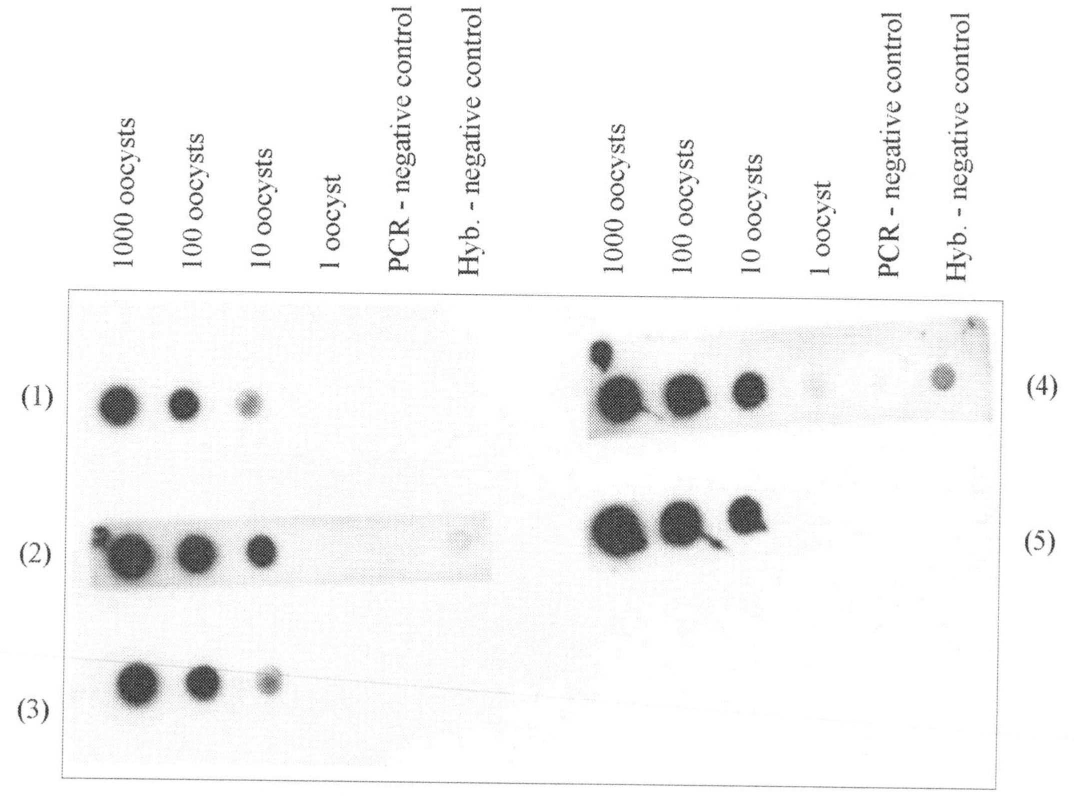


**Figure 3.5. Detection of *C. parvum* using slot blot hybridization on Nylon membrane with fluorescein-labeled DNA (1) and PNA probes (2 – 5). Hybridization was performed in: (1) conventional hybridization; (2) 1× Tris-EDTA buffer at room temperature; (3) SSC-Block-SDS buffer at room temperature; (4) 1× Tris-EDTA buffer at 37°C; (5) SSC-Block-SDS buffer at 37°C. Detection was performed using an anti-fluorescein antibody-alkaline phosphatase conjugate and CDP-Star chemiluminescent substrate.**

Dot blot with unlabeled amplicons, hydrization in SSC - Block - SDS buffer at 37° C

Southern transfer of biotin - labeled amplicons, hybridization in SSC - Block - SDS buffer at 37° C

Southern transfer of biotin - labeled amplicons, hybridization in SSC - Block - SDS buffer at 50° C

Southern transfer of biotin - labeled amplicons, hybridization in Tris - EDTA buffer at 50° C

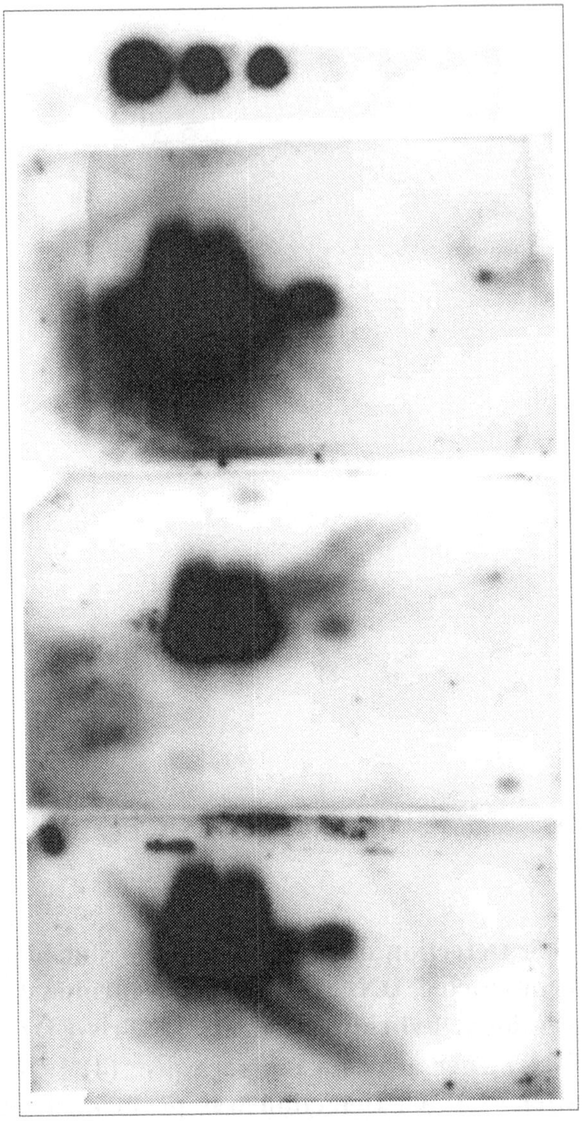

**Figure 3.6. Detection of biotin-labeled *C. parvum* amplicons using hybridization on Nylon membrane and the PNA probe. Detection was performed using anti-fluorescein antibody and CDP-Star chemiluminescent substrate. Each membrane shows oocyst concentrations in PCR of 100,000, 10,000, 1,000, 100 and 10 oocysts (from left to right) followed by PCR negative control.**

PNA probes were also effective when hybridized in microplates, although the hybridization buffer required was different (100 mM Tris and 100 mM NaCl) than with the DNA probes. As a result, two different buffers were needed for the assay: the SSC-SDS-blocking reagent buffer for biotin-labeled amplicon pre-coating to the microplate, and the Tris-NaCl buffer for the hybridization.

Attempts were made to decrease the incubation period from 1 hour at 37°C, to 30 min at room temperature or 37°C. An assay sensitivity of less than 5 oocysts (number of oocysts added into PCR, lowest concentration tested) was achieved with no observable difference in signal intensity between the two hybridization temperatures evaluated, although results were less variable at 37°C than at room temperature. When the microplate hybridization was conducted at room temperature, the probe concentration (0.10, 1.0 or 10 nM) did not significantly affect the signal intensity, but the strongest signal was obtained with the highest probe concentration when the hybridization was conducted at 37°C.

These results suggest that in this particular case (hybridization of *C. parvum* in microplates) and in contrast to the results obtained for conventional membrane hybridization, the PNA probes do not significantly improve the assay sensitivity compared with DNA probes. Although increasing hybridization temperature is necessary to obtain sensitive results, this step increases the complexity of the assay and is against the purpose of using PNA probes. Further development by Boston Probes, Inc., indicated that probes shorter than 15 bp are less stable than longer probes (the probe stability is even more critical for the rapid screening assay which uses microplate hybridization instead of membrane hybridization). In addition, because PNA probes tend to precipitate, they have to be resuspended in 50% N,N'-dimethylformamide, a hazardous compound that needs to be handled with great care. Also, the need for two different buffers (one for the amplicon pre-coating and one for the microplate hybridization) somewhat complicates the assay. Therefore, under the conditions tested in this study, PNA probes did not markedly outperform DNA probes, and were not further examined.

The other advantages of the PNA probe format (better performance in *in-situ* hybridization, identification of point mutation, resistance to enzymatic degradation, etc) although very important do not apply to the rapid screening assay. In addition, labeled PNA probes are currently about twice the cost of their DNA counterparts, which does not meet the low-cost criteria of the assay, and are more difficult to obtain as very few manufacturers are producing them. Improvements in assay conditions and eventually lower cost of PNA probes may make them more attractive at a later date.

**Optimization of End-Point Colorimetric Detection**

Although a microplate reader is not necessary for evaluating results obtained with the colorimetric detection, an alternative approach was evaluated. The absorbance of the colorimetric signal was read using the microplate reader at a wavelength of 550 nm for some experiments. This mode of characterizing the results was used to provide relative quantification of the results and to better evaluate the pale signals.

Different reaction times for the NBT/BCIP colorimetric substrate were evaluated (15 – 180 min and overnight). A period of 2 hours at room temperature was selected. Color develops until approximately 2 hours, but longer incubation periods lead to non-specific background.

**Assay Sensitivity**

Using the optimized amplification and microplate hybridization conditions, a sensitivity of 10 oocysts was achieved (Figure 3.7). Figure 3.7 presents results of absorbance measured at a wavelength of 550 nm with the microplate reader while Table 3.13 displays the results of visual evaluation. Using the microplate reader helps better characterize the pale signal, however further work is needed to evaluate the exact wavelength of the NBT/BCIP substrate, possible interferences, and acceptance criteria for positive and negative values.

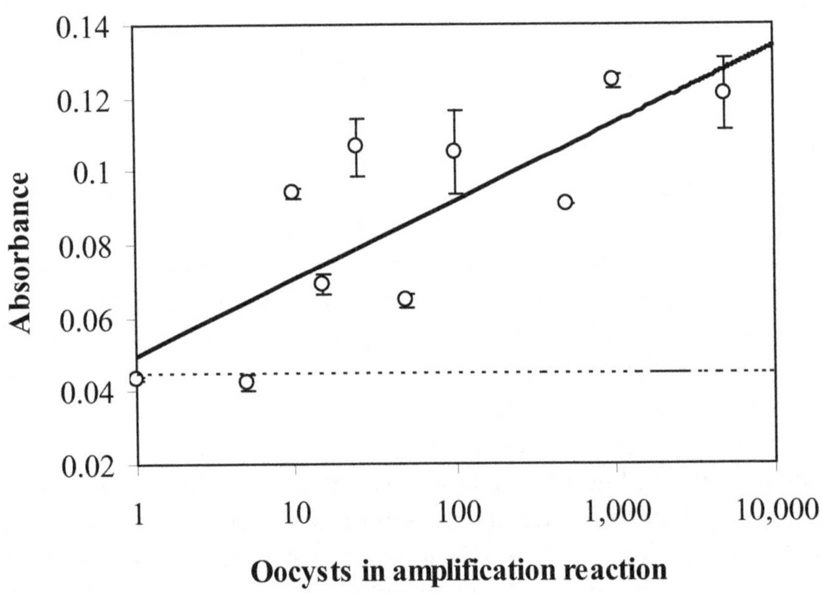


**Figure 3.7. Sensitivity of the rapid screening assay for detection of *C. parvum* oocysts. Error bars are the standard deviation of duplicate experiments. The dashed line is the detection threshold for this particular experiment based on the maximum absorbance ($A_{550}$) of the PCR and hybridization negative controls.**

**Table 3.13**
**Summary of signal intensity obtained with *C. parvum* using Approach ACM/SL-PRB and colorimetric detection**

| Number of Oocysts in PCR | Visual Results |
|---|---|
| 5,000 | +++ |
| 1,000 | +++ |
| 500 | +++ |
| 100 | +++ |
| 50 | ++ |
| 25 | ++ |
| 15 | ++ |
| 10 | + |
| 5 | - |
| 1 | - |
| PCR negative control | - |
| Hybridization negative control | - |

Note: -, no signal; +/-, weak signal; + to ++++, pale to darkest signal.

**Additional Issues With the ACM/SL-PRB Approach for Detection of *Cryptosporidium parvum***

As with the PCM approach the critical steps in the ACM/SL-PRB approach for detection of *C. parvum* were: amplicon denaturation and pre-coating techniques; sample volume; SSC and probe concentrations; and sufficient contact with the NBT/BCIP colorimetric substrate. Other aspects of the assay that were examined but did not significantly influence the signal intensity included the primer concentration, type of polymerase enzyme, dUTP/UNG instead of dTTP, SDS and blocking reagent concentrations, and hybridization period.

***Optimization of Amplification Conditions***

Because it was suspected that unincorporated biotin-labeled primers and primer dimers might occupy some microplate streptavidin sites that would then be unavailable for amplicon attachment to the microplate, a PCR amplicon purification step was included. Gel electrophoresis of purified and non-purified amplicons, and slot blots on Nylon membrane were used to evaluate the presence of unincorporated primers and primer dimers and select the most appropriate system. Results indicated that the Wizard PCR Preps DNA Purification System was effective for the removal of unincorporated biotin-labeled primers and did not impair the stability of the biotin label. However, later experiments indicated that this step was not necessary as sufficient sensitivity was achieved without it. In fact, there was some loss of amplicons during the purification process. It is possible that the purification step effectively improves the signal intensity by removing unincorporated biotin-labeled primers and primer dimers, but also removes amplicons at the same time, leading to a nil overall effect (Figure 3.3). Subsequent experiments conducted with different concentrations of the positive control (biotin-AP at 0.1 – 1%, v/v) suggest that there was an excess of streptavidin binding sites relative to the amount of biotin-labeled products introduced in the microplate wells.

### *Optimization of Microplate Hybridization Conditions*

The amplicon denaturation process adopted was incubation at 95°C for 5 min followed by 99°C for 1 min, after which the amplicons were held on crushed ice until added to the microplate. The denatured biotin-labeled amplicons were introduced into the microplate wells to attach to the streptavidin coating, which raises the possibility of amplicon re-annealing to itself during this pre-coating period and thus not being available for probe hybridization. To improve the stability of the denatured amplicons, it was proposed to use formamide at high temperature to denature the amplicons. The denaturation temperature is decreased by 0.65°C from the melting temperature ($T_m$, temperature at which 50% of the DNA is disassociated) for each 1% formamide used (Brown, 1991). As a result, a temperature of 50.7°C would be sufficient to denature *C. parvum* amplicons generated with the CPHSP2423F and CPHSP2764R primers ($T_m$ of 83.2°C) in 50% formamide. Another consideration is that the polystyrene microplates such as those used in this assay cannot remain in contact with more than 50% formamide for 1 hour at 56°C. Considering that a low salt concentration also helps amplicon denaturation, attempts were made to denature the amplicons in deionized water and formamide at 50°C. Denaturation at high pH such as used during membrane hybridization was also attempted, although this practice is not recommended, as it is possible that the peptide bond of the biotin may be hydrolyzed under alkaline conditions. Experiments indicated that the heat-denaturation process was sufficient, with a microplate pre-coating at 4°C instead of room temperature to limit amplicon re-annealing to itself during the pre-coating period. The same buffer was used for microplate pre-coating and hybridization, and contained SSC, blocking reagent, and SDS, as selected for Approach PCM/ML-dUTP.

### *Assay Duration*

The detection of *C. parvum* using Approach ACM/SL-PRB and the colorimetric detection system indicated that the optimized assay is relatively unaffected by changes in the parameters tested. Some parameters such as amplicon denaturation and microplate pre-coating techniques are important, but it seems that others do not really influence the assay sensitivity and specificity (example reagent concentrations and incubation periods). Therefore, two experiments were conducted with the same biotin-labeled amplicons, and amplification and hybridization reagents: (1) with the optimized hybridization conditions, and (2) with reduced incubation periods (Table 3.14). Each sample was processed in duplicate (two microplate wells), and the results obtained are presented in Table 3.15 and Figure 3.8. The sensitivity obtained during these particular assays was weak, only 50 oocysts for both the standard and short methods. However results indicated that the incubation periods could be decreased without impairing the assay sensitivity, which makes this assay attractive as a rapid screening tool. Further experiments with even shorter incubation periods would be necessary to determine the minimum possible assay duration. The similarity of the results obtained by both duplicate demonstrates the reproducibility of the assay.

**Table 3.14**
**Assay incubation periods (min) for the standard and the short assays**

| Assay steps | Standard method | Short method |
|---|---|---|
| Microplate pre-washing (including rinses) | 15 | 6 |
| Microplate coating with biotin-amplicons | 60 | 30 |
| Post pre-coating washes (x2) | 20 | 10 |
| Microplate hybridization | 180 | 105 |
| Post-hybridization washes (x2) | 20 | 10 |
| Antibody reaction | 30 | 20 |
| Post-antibody washes (x2) | 10 | 4 |
| Colorimetric detection | 120 | 120 |
| Post-detection washes (x2) | 10 | 4 |
| Total incubation time | 7.75 hours | 5.15 hours |

Only the incubation periods that were modified are presented.

**Table 3.15**
**Summary of signal intensity obtained with the standard and short assay incubation periods**

| Number of oocysts | Standard Method | | Short Method | |
|---|---|---|---|---|
| in PCR | Duplicate 1 | Duplicate 2 | Duplicate 1 | Duplicate 2 |
| 5,000 | +++ | +++ | +++ | +++ |
| 1,000 | +++ | +++ | +++ | +++ |
| 500 | +++ | +++ | +++ | +++ |
| 100 | ++ | + | ++ | + |
| 80 | + | ++ | + | ++ |
| 50 | ++ | + | ++ | + |
| 25 | - | - | - | - |
| 15 | - | - | - | - |
| 10 | - | - | - | - |
| PCR negative control | - | - | - | - |

Note: -, no signal; +/-, weak signal; + to ++++, pale to darkest signal.

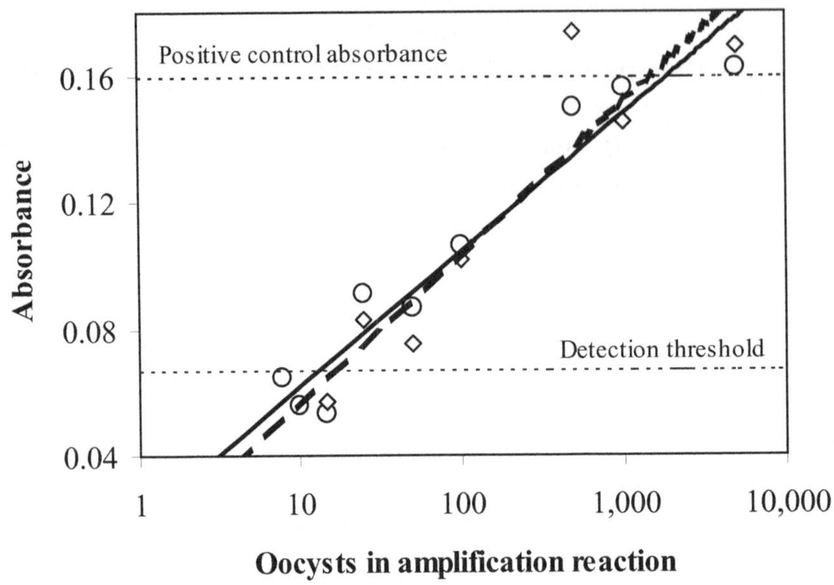


**Figure 3.8. Comparison of standard and modified rapid screening assays for the detection of *C. parvum* oocysts. The duration of the microplate hybridization and detection portion of the standard assay was 7.8 h (○, ——) compared to 5.2 h for the shortened version (◊, — —). Each datapoint is the mean of duplicate microplate wells. The detection threshold is the maximum absorbance value ($A_{550}$) of negative PCR and hybridization controls.**

## DETECTION OF ENTEROVIRUSES

The rapid screening assay was also evaluated with echovirus 1 as a representative enterovirus, a major group of pathogens with RNA genomes. When using Approaches PCM/ML-dUTP and ACM/SL-PRB for the detection of echovirus 1, successful results were obtained with amplification and hybridization conditions similar to those used for *C. parvum* detection, with very little optimization necessary. All results presented in this section were obtained with the colorimetric detection.

### Approach PCM/ML-dUTP Used for the Detection of Enteroviruses

Selected amplification and hybridization conditions were tested and are presented in Table 3.16. Only single PCRs were performed with the enteroviruses. Polymerization was conducted with higher overall dUTP concentration (250 μM each dATP, dCTP and dGTP, with 500 μM dUTP) as optimized previously for *C. parvum* detection. Results indicated no observable differences between the ratios of DIG-11-dUTP:dUTP tested (Table 3.17). For development purposes, a ratio of 1:4 was selected (same ratio as used for *C. parvum* detection).

With the enterovirus system, experiments indicated the need for a much higher probe concentration (5 nM) than that used with *C. parvum* (40 pM), which suggested less efficient incorporation of biotin into the enterovirus probe. Dot blots of the enterovirus biotin-labeled probes were compared to the *C. parvum* biotin-labeled probe, and results indicated that the biotin concentration was more than 10 times higher on the *C. parvum* probe than on the enterovirus probes (Figure 3.9). Results indicated that 6X SSC (the same concentration as determined with *C. parvum*) and 5.0 nM probe concentrations provided the strongest signal with less non-specific colorimetric detection, although the difference between conditions was not pronounced (Table 3.17). In general, a sensitivity of 0.12 PFU (approximately 120 virions) was easily achieved with Approach PCM/ML-dUTP and the colorimetric detection for Echovirus 1.

**Table 3.16**
**Amplification and hybridization conditions evaluated for Echovirus 1 detection using Approach PCM/ML-dUTP**

| Conditions | Range tested | Optimum or Selected conditions |
|---|---|---|
| Ratio of DIG-11-dUTP:dUTP | 1:4 – 1:10 | 1:4 |
| Probe concentration | 40 pM – 5 nM | 5 nM |
| SSC concentration | 3X – 10X | 6X |
| Blocking reagent concentration | 1 – 3% | 1% |
| SDS concentration | 0.02% | 0.02% |
| Hybridization period (hours at 55°C) | 2 – 4 h | 2 – 3 h |

**Table 3.17**
**Summary of signal intensity obtained with different DIG-11-dUTP:dUTP ratios and SSC concentrations**

| PFU of Echovirus 1 | 1:4 DIG-11-dUTP:dUTP | | | 1:6 DIG-11-dUTP:dUTP | | |
|---|---|---|---|---|---|---|
| in PCR | 4X SSC | 6X SSC | 8X SSC | 4X SSC | 6X SSC | 8X SSC |
| 1,200 | +++ | +++ | +++ | + | + | + |
| 120 | +++ | ++++ | ++++ | +++ | +++ | +++ |
| 12 | +++ | ++ | ++ | ++ | ++ | ++ |
| 1.2 | +/- | +/- | +/- | +/- | +/- | +/- |
| 0.12 | - | - | - | - | - | - |
| Reverse transcriptase negative control | - | - | - | - | - | - |
| PCR negative control | - | - | - | - | - | - |

Note: -, no signal; +/-, weak signal; + to ++++, pale to darkest signal.

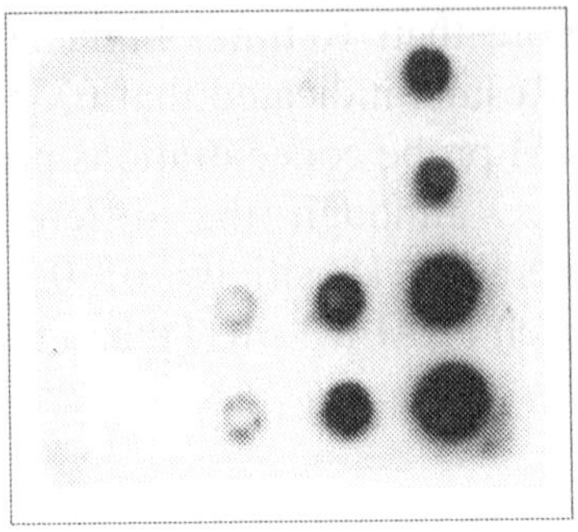


**Figure 3.9. Signal intensity of various biotin-labeled probe concentrations using a dot blot on Nylon membrane for the PanEV-1 probe targeting enteroviruses and the CPHSP2475 probe specific for *C. parvum*. Detection was performed using streptavidin-alkaline phosphatase conjugate solution followed by CDP-Star chemiluminescent substrate.**

### Approach ACM/SL-PRB Used for the Detection of Enteroviruses

When evaluating Approach ACM/SL-PRB for the detection of enteroviruses, both the forward and reverse primers were biotin labeled, and only single PCRs were performed. Results indicated that the rapid screening assay and the enterovirus system are only slightly affected by changes in amplification and hybridization conditions. When working with *C. parvum* and Approach ACM/SL-PRB, the primer concentration was decreased from 500 to 250 nM to limit attachment of unincorporated biotin-labeled primers to the microplate streptavidin, which could limit the attachment of the biotin-labeled amplicons. In the case of the enteroviruses, this modification was not necessary. Sensitive results (less than 1 PFU) were obtained with the same primer concentration (600 nM) for both Approaches PCM/ML-dUTP and ACM/SL-PRB. The different microplate hybridization conditions tested are presented in Table 3.18.

Results indicated that the higher salt and probe concentrations used during microplate hybridization had only a marginal effect on the colorimetric signal while the blocking reagent concentration had no effect. It was not necessary to further optimize the Echovirus 1 assay as good sensitivity (strong signal) was obtained with most conditions tested. The same conditions as determined for *C. parvum* detection were therefore used for the echovirus 1 assay as well.

DIG- and fluorescein-labeled PanEV-1 probe were compared for echovirus 1 detection using Approach ACM/SL-PRB. The colorimetric detection was performed with anti-DIG-AP or anti-fluorescein-AP antibody and NBT/BCIP substrate. Results indicated slightly stronger signal intensity with the fluorescein-labeled probe than with the DIG-labeled probe. The assay sensitivity was less than 0.12 PFU (approximately 120 virions, the minimum concentration tested), with no significant increase in signal strength with increasing Echovirus 1 concentration. When conducting similar comparison using Approach ACM/SL-PRB and *C. parvum*, results were more sensitive and reproducible with the DIG hapten than with the fluorescein hapten. Therefore, a DIG-labeled probe was used for further experiments considering the strong signal intensity obtained with both probe hapten in the case of Echovirus 1.

**Table 3.18**
**Hybridization conditions evaluated for echovirus 1 detection using Approach ACM/SL-PRB**

| Hybridization conditions | Range tested | Optimum or selected conditions |
|---|---|---|
| Probe concentration | 0.2 – 1 nM | 500 pM |
| SSC concentration | 4X – 10X | 6X |
| SDS concentration | 0.02% | 0.02% |
| Blocking reagent concentration | 1 – 3% | 1% |
| Hybridization period (55°C) | 2 – 3.75 h | 2 – 3 h |

**Assay Sensitivity**

The sensitivity was thoroughly evaluated using fresh stock solutions of Echovirus 1, and the lowest concentrations detected were 0.016 and 0.005 PFU (approximately 16 and 5 virions, respectively; Figure 3.10). Visual evaluation indicated that positive results (+) were observed for Echovirus 1 concentrations of 0.05 PFU and above, and weak or unclear results (+/-) for concentrations of 0.01 PFU and below.

**Assay Specificity**

Studies conducted by De Leon et al. (1990) indicated that the primers and probes selected recognize over 40 human enteroviruses, but do not amplify nucleic acid from bovine enteroviruses 1 to 7, porcine enteroviruses 1 to 8a, 8b or 8c, Hepatitis A virus, Reovirus, Norwalk virus, Rotavirus, nor Astrovirus.

As a verification of specificity, the rapid screening assay specificity was evaluated using Approach ACM/SL-PRB for the detection of enteroviruses with mammalian cells (BGMK, Caco-2, and HCT-8), protozoa (*C. parvum*, *C. muris* and *G. duodenalis*), bacteria (*E. coli* O157:H7, *Salmonella typhimurium*, *Clostridium perfringens*, *Yersenia enterocolitica*, *Aeromonas hydrophila*, *Shigella sonnie*, *Enterobacter cloacae*, *Citrobacter freundi*, *Pseudomonas fluorescens*, and *Klebsiella pneumoniae*), and a yeast (*Saccharomyces cerevisiae*). Different concentrations of Echovirus 1 were included as positive controls in this assay, and results confirmed the specificity of the primers, probe and assay for the enterovirus group.

The assay specificity was also tested with other enteric viruses including Rotavirus, Hepatitis A, bovine enteroviruses (1 to 7), porcine enteroviruses (1, 3, 4, 5, 6, 7, 8, 8a, 8b, and 8c), Poliovirus 1, Coxsachie virus A9, A12, B3 and B5, and Echovirus 1 and 7 (Figure 3.11). Signal was not detected (negative results) with Rotavirus, Hepatitis A, all bovine enteroviruses, and all porcine enteroviruses. Positive signal was observed with all human enteroviruses tested: Poliovirus 1, all Coxsackievirus strains tested, and all Echovirus strains tested. These results confirm the specificity of the primers, probe and rapid screening assay for the human enteroviruses.

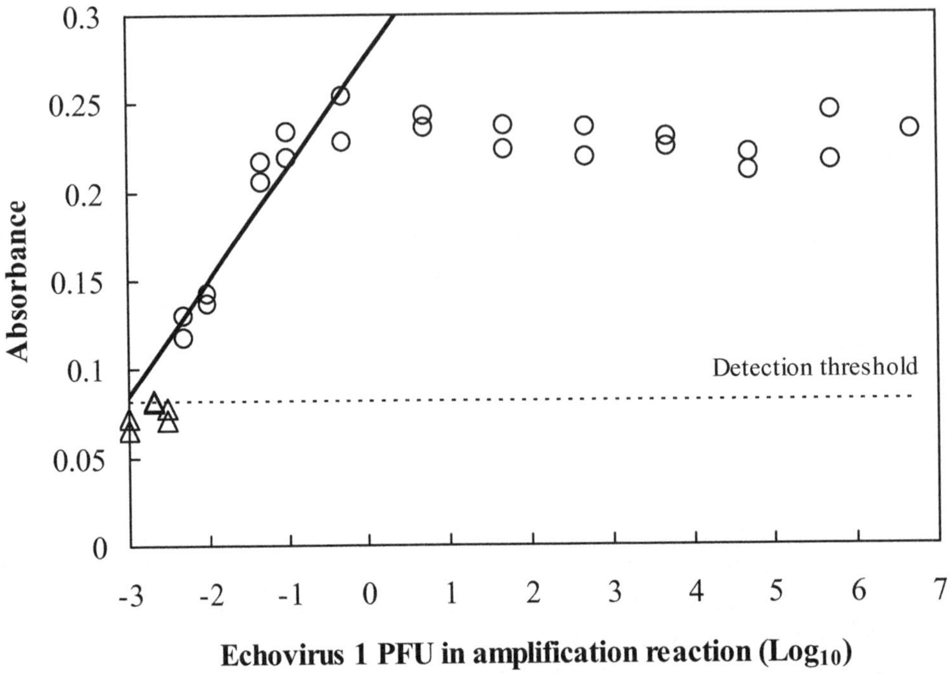


**Figure 3.10. Sensitivity of rapid screening assay with echovirus 1 (○). Duplicate wells were analyzed and absorbance was measured at 550 nm. The dashed line indicates the maximum absorbance reading obtained for the negative controls (Δ) processed during this experiment. A regression line was plotted through only the linear portion of the echovirus 1 data.**

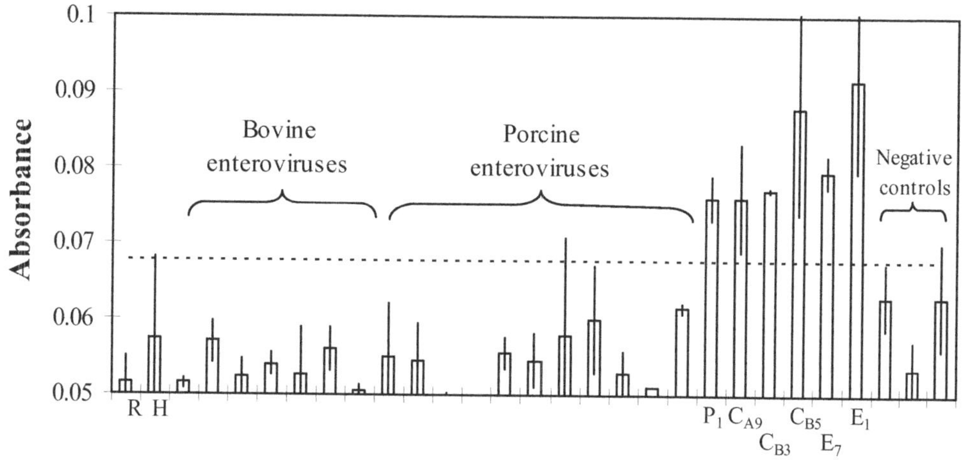


**Figure 3.11. Specificity of human pan-enterovirus specific assay utilizing amplicon-coated microplates (ACM approach). The dashed line represents the detection threshold based on the maximum absorbance for negative controls. Error bars are the standard deviation of duplicate measurements. Tested viruses were: Bovine and Porcine Enteroviruses, Rotavirus (R), Hepatitis A virus (H), Poliovirus 1 ($P_1$), Coxsackievirus A9 ($C_{A9}$), B3 ($C_{B3}$), and B5 ($C_{B5}$), and Echovirus 1 ($E_1$) and 7 ($E_7$).**

## APPLICATION OF THE ASSAY TO ADDITIONAL PATHOGENS

The transferability of the assay to other microorganisms was also assessed using *G. duodenalis* and *E. coli* O157:H7 as examples. The intent of this objective was not to re-optimize the entire assay for each microorganism, but rather to use published amplification and hybridization conditions (primers and probe sequences, PCR activity, hybridization and post-hybridization temperatures, etc.) with the optimized assay conditions (reagent concentrations, incubation periods, etc.) already determined and common to both *C. parvum* and Echovirus 1. Both Approaches PCM/ML-dUTP and ACM/SL-PRB were tested with *G. duodenalis* (two strains) and *E. coli* using the colorimetric detection.

### Approach PCM/ML-dUTP Used for the Detection of *Giardia duodenalis* and *Escherichia coli* O157:H7

When Approach PCM/ML-dUTP was used to detect *G. duodenalis* H3 isolate, less than 9 cysts (the lowest concentration tested) was successfully detected. Results indicated very little difference between the different probe concentrations tested (40 pM, 100 pM, 1.0 nM and 10 nM) which confirms that the rapid screening assay is unaffected by changes in the hybridization parameters.

Experiments indicated that the optimum microplate hybridization conditions determined for *C. parvum* and Echovirus 1 with Approach PCM/ML-dUTP and the colorimetric detection, were applicable to *E. coli* as well. The method sensitivity was however poor with 7,000 colony forming units (CFU).

**Approach ACM/SL-PRB Used for the Detection of *Giardia duodenalis* and *Escherichia coli* O157:H7**

Positive signal was obtained with *G. duodenalis* WB strain using Approach ACM/SL-PRB and the colorimetric detection. The presence of *G. duodenalis* was detected in the samples containing approximately 9 cysts (based on serial dilution of a hemacytometer enumerated stock) and no observable difference was detected between the two probe concentrations tested (1.0 and 10 nM; Figure 3.12).

When using Approach ACM/SL-PRB for the detection of *E. coli* O157:H7, the assay sensitivity was 10 CFU (compared to 7,000 CFU for the PCM approach) and no significant difference was observed between the probe concentrations tested (1.0 and 10 nM).

Results of these experiments confirm that the rapid screening assay is easily transferable to other microorganisms. They also suggest that Approach ACM/SL-PRB may be more easily transferable to other organisms as the amplification and microplate hybridization conditions do not need optimization, whereas the probe concentration may vary from one organism to another when Approach PCM/ML-dUTP is used.

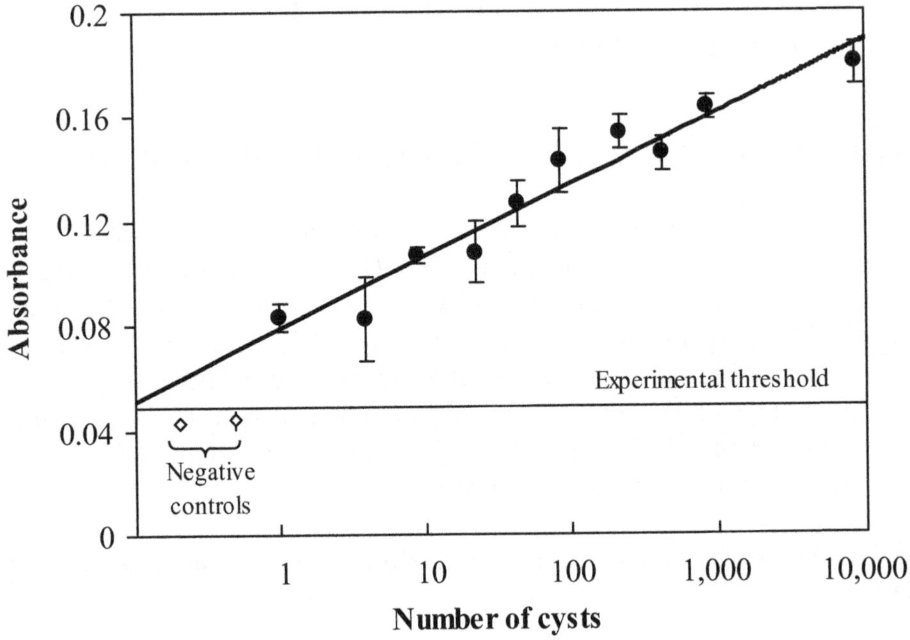


**Figure 3.12. Detection of *Giardia duodenalis* using the amplicon coated microplate (ACM) approach with a single-label DIG-probe and colorimetric detection. There was no significant difference (P = 0.88) between intensities generated with probe concentrations of 1 nM and 10 nM so the data were combined. Errors bars are standard deviations (n = 4). The detection threshold represents the maximum absorbance reading for PCR and hybridization negative controls processed during this particular experiment (n = 4 for each type of control).**

## END-POINT FLUORESCENCE DETECTION

The assay was originally developed using an AP-labeled antibody and a colorimetric substrate (NBT/BCIP) to allow a rapid and visual evaluation of the results without the need for sophisticated equipment such as a microplate reader. Other detection labels such as fluorophores (FITC, AttoPhos) or chemiluminescent substrates (CDP-Star, CSPD, and AMPPD from Tropix Inc.) are available and may improve the method sensitivity. In the case of the fluorescent detection, another potential advantage was reduced analysis time by possibly eliminating the antibody and detection substrate incubation steps, because the fluorophore is directly incorporated into the amplicons in Approach PCM/ML-dUTP or probe in Approach ACM/SL-PRB (see Figure 3.1).

### *Cryptosporidium parvum* Detection Using the Fluorescence Method

Both amplification and microplate hybridization approaches presented earlier were used to evaluate the fluorescence detection method using *C. parvum*.

#### *Approach PCM/ML-dUTP for the Fluorescent Detection of* **Cryptosporidium parvum**

Fluorescein or DIG was incorporated into the amplicons during PCR amplification using fluorescein-12-dUTP or DIG-11-dUTP, respectively. The fluorescein-labeled amplicons were used for direct fluorescent detection following microplate hybridization (no antibody and detection substrate required), whereas the DIG-labeled amplicons allowed indirect fluorescent detection following the addition of an anti-DIG-fluorescein antibody (Figure 3.1).

Results obtained with the direct detection technique were erratic, but the indirect detection provided clearer results (Figure 3.13) most likely because the antibody step amplifies the signal. At the time this experiment was conducted, the sensitivity with the indirect fluorescent detection was similar to the colorimetric detection (50 oocysts added to PCR).

It is known that fluorescein exhibits a relatively weak fluorescence compared to some of the more advanced fluorophores such as Texas Red or Rhodamine. ChromaTide Alexa Fluor 488 (Molecular Probes) possesses similar spectral characteristics to fluorescein, but is pH insensitive, significantly brighter and much more photostable. During one experiment conducted with *C. parvum* and Approach PCM/ML-dUTP, the direct detection of fluorescein-12-dUTP or ChromaTide Alexa Fluor 488-5-dUTP incorporated into the amplicons (at different ratio of labeled-dUTP:dUTP) was compared with the indirect fluorescent detection (DIG-11-dUTP incorporated in the amplicons and anti-DIG-fluorescein antibody). Fluorescence was read after microplate washing using different buffers. Results with the direct detection technique were inconsistent and seemed to depend on the washing buffer used. In general, the indirect detection was less variable, and provided much stronger signal than the direct detection, most likely because of signal amplification resulting from the antibody reaction. Also, the indirect fluorescein signal was stronger than the direct signal obtained with ChromaTide Alexa Fluor. It is possible that the smaller fluorescein molecule was better incorporated into the amplicons than the ChromaTide Alexa Fluor molecule. These results suggest that the fluorescence detection cannot provide a significant improvement over the colorimetric detection, as the method sensitivity was 50 *C. parvum* oocysts in both cases.

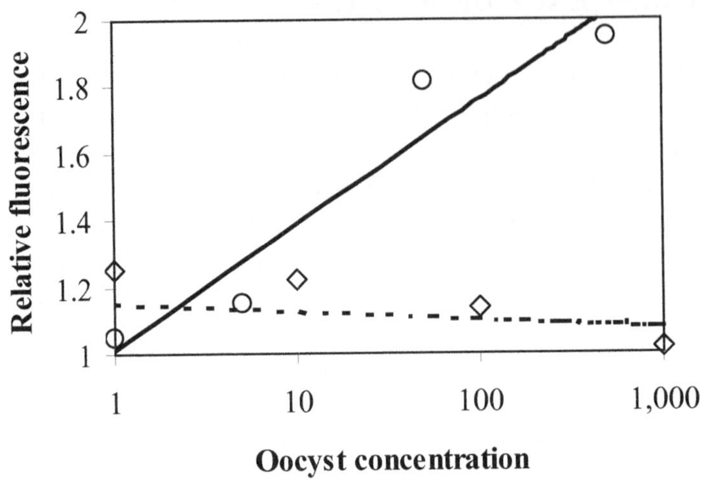


**Figure 3.13. Comparison of direct (◊, dashed line) and indirect (○, solid line) fluorescence detection of *C. parvum* using the probe-coated microplate approach (PCM/ML-dUTP). Fluorescence values were calculated relative to the maximum fluorescence of the negative controls for each type of assay (relative fluorescence for negative controls = 1).**

### *Approach ACM/SL-PRB for the Fluorescent Detection of* **Cryptosporidium parvum**

As observed with Approach PCM/ML-dUTP, the direct detection of the probe-amplicon hybrids (fluorescein-labeled probe) provided erratic results, even when high probe concentrations were used (10 and 100 nM). With the indirect detection (DIG-labeled probe and anti-DIG-fluorescein antibody), the best sensitivity observed was 10 oocysts (concentrations added to PCR), and stronger signal was obtained with a reading buffer consisting of 1X Tris-EDTA than with a buffer containing Tris and 1% (w/v) BSA (Tris-BSA buffer), although the trends observed were similar (Figure 3.14). The two antibody concentrations tested (0.2 and 2 μg/mL) during the indirect detection did not really affect the signal intensity.

### Echovirus 1 Detection Using the Fluorescence Method

The enterovirus assay was also used to evaluate the fluorescence detection method. Both amplification and microplate hybridization approaches presented earlier were used to evaluate the fluorescence detection method using echovirus1.

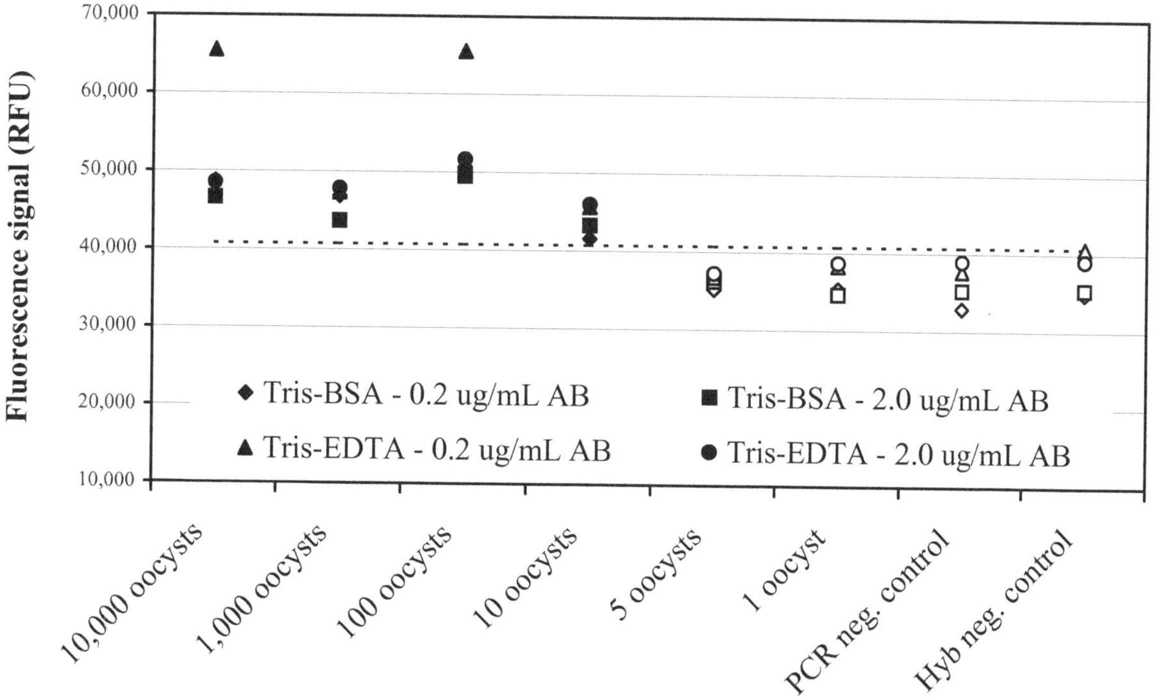


**Figure 3.14. Fluorescent signal obtained during the indirect detection of *C. parvum* using Approach ACM/SL-PRB. Two antibody concentrations (0.2 and 2.0 µg/mL) and two reading buffers were tested (Tris with 1% BSA and 1X Tris-EDTA). The dotted line indicates the maximum fluorescence reading obtained for the negative controls and the indirect detection processed during this experiment. The full symbols represent positive signals (absorbance value above the negative controls analyzed), and the open symbols negative signals.**

### *Approach PCM/ML-dUTP for the Fluorescent Detection of Enteroviruses*

As observed during the experiments involving *C. parvum* fluorescent detection, results obtained from the direct detection of fluorescein-labeled amplicons (FITC-12-dUTP incorporation during PCR amplification, without antibody nor substrate following microplate hybridization) did not provide coherent signals and the readings and trends observed were very dependent on the reading buffer used.

The indirect detection (DIG-11-dUTP during PCR, with anti-DIG-fluorescein at the antibody incubation step) provided adequate signal amplification and probe-amplicon hybrid detection. Results suggested that the higher antibody concentration (2.0 µg/mL instead of 0.20 µg/mL) provided stronger signal (Figure 3.15).

Although the literature gives excitation and emission wavelengths of 494 and 520 nm, respectively, for fluorescein, a lower excitation wavelength is often used (450-490 nm). As a result, an excitation wavelength of 480 nm was investigated with Echovirus 1 and Approach PCM/ML-dUTP. Readings were however more consistent when conducted at 495 nm (Figure 3.16), which was the wavelength used for all prior and subsequent experiments.

Considering the results obtained when investigating the fluorescent detection technique, we recommend using the more practical, visual colorimetric detection. Overall, the fluorescent detection did not improve the assay sensitivity with either approaches or microorganisms examined, and requires sophisticated equipment (a microplate reader with appropriate excitation and emission filters) which complicates and increases the cost of the assay. Also, the direct detection gave erratic results, and the indirect detection is necessary which does not decrease the assay processing time as it was originally anticipated. However, results indicated that it is possible to use the fluorescence detection technique if necessary, which may be the case when evaluating particular microorganisms or conditions.

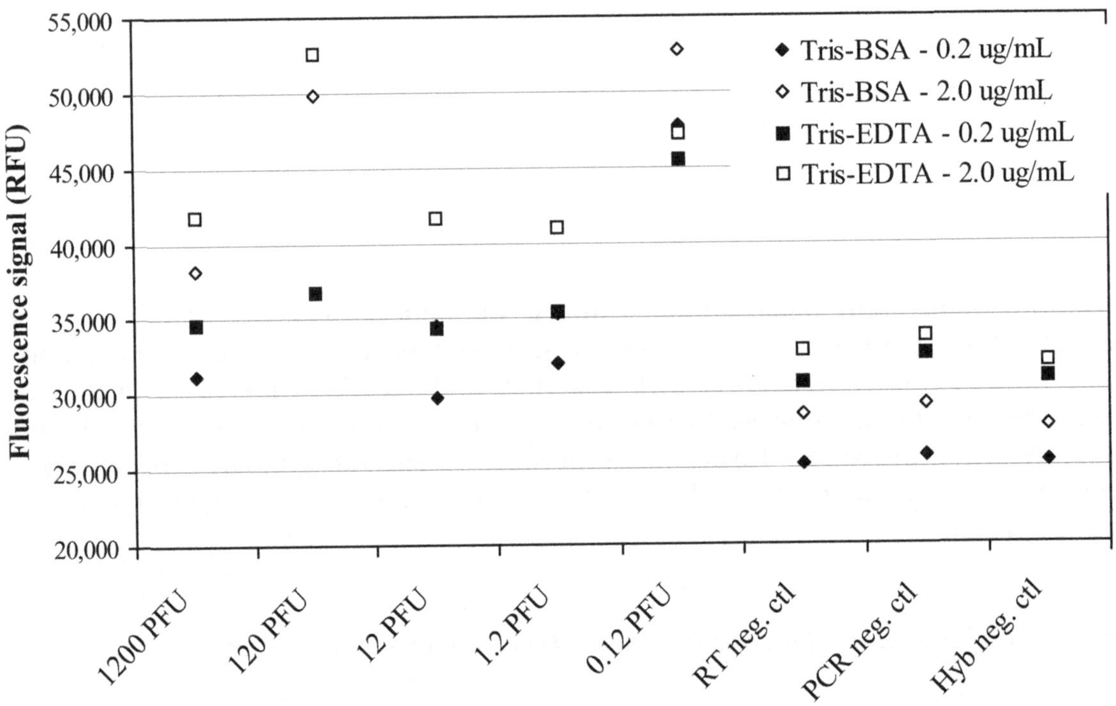


**Figure 3.15. Fluorescent signal obtained during the indirect detection of Echovirus 1 using Approach PCM/ML-dUTP. Two reading buffers were tested (Tris with 1% BSA and 1X Tris-EDTA) and Two Antibody Concentrations (0.2 and 2.0 μg/mL).**

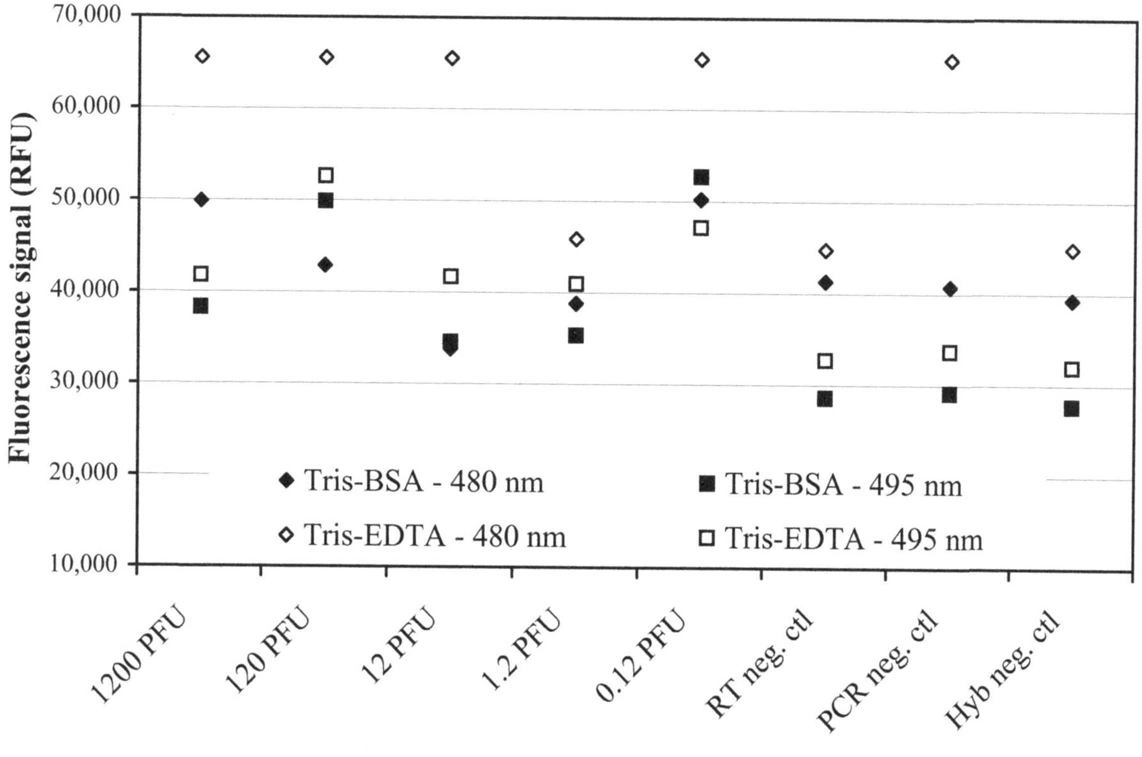


**Figure 3.16. Effect of excitation wavelength on fluorescent signal obtained during the indirect detection of Echovirus 1 using Approach PCM/ML-dUTP. The antibody concentration was 2.0 µg/mL and two reading wavelengths were tested (480 and 495 nM).**

## COMPARISON OF APPROACHES

Although Approaches PCM/ML-dUTP and ACM/SL-PRB provided similar results in terms of sensitivity and specificity, the preferred approach was ACM/SL-PRB with the colorimetric detection because of its lower cost, shorter assay duration, verification of amplicons by gel electrophoresis if necessary, sensitivity and transferability to other microorganisms.

### Cost

Approach ACM/SL-PRB is less costly than Approach PCM/ML-dUTP, as presented in Table 3.19. These values were determined assuming 50-µL PCR amplification, and 72 samples per microplate with 24 negative or positive controls (for a total of 96 well per microplate). They include costs of materials and supplies such as microplate, labeled primers and probe, sterile pipet tips and all reagents, but do not include equipment costs such as thermal cycler and hybridization oven. The high cost of the DIG-11-dUTP used in Approach PCM/ML-dUTP ($9.42 per 50-µL PCR) is responsible for the higher cost of this approach.

**Table 3.19**
**Costs of PCR amplification and microplate hybridization associated with Approaches PCM/ML-dUTP and ACM/SL-PRB**

| | Approach PCM/ML-dUTP | Approach ACM/SL-PRB |
|---|---|---|
| PCR amplification | $11.25 per reaction<br>$810.18 per microplate | $1.97 per reaction<br>$141.58 per microplate |
| Microplate hybridization | $0.82 per well<br>$79.10 per microplate | $0.82 per well<br>$78.99 per microplate |
| Total | $889.28 per 96-well microplate | $220.57 per 96-well microplate |

### Visualization of Results

The biotin-labeled amplicons obtained with Approach ACM/SL-PRB can be seen on gel electrophoresis, whereas DIG-labeled amplicons obtained in Approach PCM/ML-dUTP have not been observed in standard ethidium bromide stained gels. The ability to visualize amplicons is important as it provides an additional verification step throughout the process.

The multiple experiments conducted during this study indicated that the samples that showed a positive signal in gel electrophoresis were also positive by the rapid screening assay, and those that were negative by gel electrophoresis were negative by the assay. The rapid screening assay can thus be used as a robust confirmation procedure for the presence of specific microorganisms in natural water.

### Assay Duration

The assay duration is similar for both approaches: 10.3 h and 10.8 h for Approaches PCM/ML-dUTP and ACM/SL-PRB, respectively. Details of the incubation periods necessary for each approach are presented in Table 3.20, considering analysis of a full 96-well microplate. These estimates do not include time necessary for reagent preparation nor PCR amplification, which are similar for both approaches.

The assay durations are very conservative values as these assays were designed to provide sensitive, specific and reliable results during method development. However, the time required to perform several steps can easily be reduced, as presented earlier for the detection of *C. parvum* using Approach ACM/SL-PRB and the colorimetric detection. Although the short method significantly decreased the assay duration (from 7.75 hours to 5.15 hours in that particular experiment which involved only 30 samples per microplate), its duration could be further decreased. As an example, the hybridization period could be reduced, and the antibody reaction could be limited to 10 min. All microplate incubations conducted after adding wash buffers could be eliminated (currently 5 to 10 min).

**Table 3.20**
**Approximate duration of assays (Approaches PCM/ML-dUTP and ACM/SL-PRB)**

| Approach PCM/ML-dUTP | | Approach ACM/SL-PRB | |
|---|---|---|---|
| Operations | Duration (min) | Operations | Duration (min) |
| Pre-wash microplate | 60 | Pre-wash microplate | 60 |
| Add pre-coating buffer | 15 | Add pre-coating buffer | 15 |
| | | Load biotin-amplicons in microplate | 60 |
| Incubate at room temperature | 30 | Incubate at 4°C | 60 |
| Microplate washes (x2) | 50 | Microplate washes (x2) | 50 |
| Add hybridization buffer | 15 | Add hybridization buffer | 15 |
| Load DIG-amplicons into microplate | 60 | | |
| Incubate at hybridization temperature | 120 | Incubate at hybridization temperature | 120 |
| Microplate washes (x2) | 50 | Microplate washes (x2) | 50 |
| Add antibody (anti-DIG-AP) | 15 | Add antibody (anti-DIG-AP) | 15 |
| Incubate at room temperature | 30 | Incubate at room temperature | 30 |
| Microplate washes | 40 | Microplate washes | 40 |
| Add NBT/BCIP | 15 | Add NBT/BCIP | 15 |
| Incubate at room temperature | 120 | Incubate at room temperature | 120 |
| Total duration: | 620 | Total duration: | 650 |
| Attended | 260 | Attended | 260 |
| Non-attended | 360 | Non-attended | 390 |

**Assay Sensitivity**

The assay sensitivities for both approaches were the following: less than 5 *C. parvum* oocysts or 0.12 Echovirus 1 PFU with Approach PCM/ML-dUTP, and 10 *C. parvum* oocysts or 0.005 to 0.016 Echovirus 1 PFU with Approach ACM/SL-PRB. In the case of *C. parvum*, sensitivities of 5 to 10 oocysts are similar considering that concentrations were evaluated using serial dilutions of hemocytometer counts (counts less than 10 oocysts are considered imprecise). In the case of enteroviruses, the sensitivity obtained with the rapid screening assay was much better than with plaque forming assays (1 to 10 PFU). It may appear that a better sensitivity was obtained with Approach ACM/SL-PRB. However, sensitivity analyses conducted with this group of microorganisms and Approach PCM/ML-dUTP were limited because at the time these experiments were conducted, it was realized that Approach ACM/SL-PRB was superior for the reasons presented above.

**Assay Transferability to Other Microorganisms**

Experiments conducted with both approaches used for the detection of the four proposed microorganisms indicate that the probe concentration may need to be reviewed when Approach PCM/ML-dUTP is used as different biotin concentrations were observed on the probes. In the case of Approach ACM/SL-PRB, the probe carries the detection label (DIG), which does not seem to influence the assay sensitivity as an amplification step (the antibody reaction) follows probe hybridization. The need for probe concentration optimization when the assay is transferred to another microorganism does not make Approach PCM/ML-dUTP practical.

## EVALUATION OF THE RAPID SCREENING ASSAY WITH SEEDED ENVIRONMENTAL WATER SAMPLES

This aspect of the rapid screening assay evaluation included both *C. parvum* and echovirus 1. Target microorganisms were seeded in natural waters, recovered using standard or accepted techniques, and detected using both conventional methods and the rapid screening assay (Approach ACM/SL-PRB) with the colorimetric detection. Oocysts of *C. parvum* were seeded in surface water as this microorganism is mainly found in surface water, whereas echovirus 1 was seeded into ground water. The goal of the rapid screening assay was not to develop a microorganism concentration or extraction technique, but rather a detection method. Therefore, microorganism concentration, recovery and nucleic acid extraction techniques were not evaluated in this study.

### Evaluation of the Rapid Screening Assay Using *Cryptosporidium parvum* in Water Samples

In a preliminary experiment, 500 oocysts were seeded into two 10-L carboys containing untreated source water from the influent of Metropolitan's F.E. Weymouth Filtration Plant (La Verne, California). The background level of *C. parvum* was not measured during assay evaluation but the running annual average at Weymouth plant influent is 0.003 oocysts/L based on in-house monthly monitoring. *Cryptosporidium* was detected only once in untreated influent in 2 years of monitoring at a concentration of 1 oocyst in 10 L. The seeded oocysts were recovered following Method 1623. The 500-µL concentrate obtained was split in two fractions for analysis by:

1. Staining and enumeration by IFA in duplicate aliquots of 100 µL.
2. DNA extraction and oocyst detection using the rapid screening assay (Approach ACM/SL-PRB and colorimetric detection). DNA extraction was performed on aliquots of 100 and 200 µL of concentrates. Each fraction was processed in duplicate.

Results obtained with the IFA technique indicated that 150 oocysts were recovered in the 500-µL concentrate, which represents a recovery efficiency of 30%. With the rapid screening assay, all aliquots were tested in duplicate and provided a positive signal, indicating that the rapid screening assay was able to detect samples containing as few as 3 oocysts (based on IFA enumeration of parallel samples). A repeat of this experiment with different concentrations of oocysts seeded into treatment plant influent water demonstrated that positive signal was obtained with a DNA extract containing the equivalent of 2 oocysts (Table 3.21). Therefore, the rapid screening assay has a very good sensitivity with microorganisms recovered from natural water.

**Table 3.21**
**Detection of *C. parvum* seeded in natural surface water**

| Volume of oocyst concentrate (μL) | Dilution | Average number of oocysts in PCR* | n | Absorbance at 550 nm (mean ± s.d.) |
|---|---|---|---|---|
| 200 | No | 18 | 4 | 0.146 ± 0.008 |
| 200 | 1:2 | 9 | 4 | 0.142 ± 0.028 |
| 200 | 1:4 | 4 | 4 | 0.113 ± 0.046 |
| 100 | No | 9 | 4 | 0.160 ± 0.017 |
| 100 | 1:2 | 4 | 4 | 0.164 ± 0.005 |
| 100 | 1:4 | 2 | 4 | 0.110 ± 0.048 |
| PCR positive control | | | 4 | 0.172 ± 0.006 |
| PCR negative control | | | 2 | 0.070 ± 0.009 |
| Hybridization negative control | | | 2 | 0.061 ± 0.005 |

* Based on IFA enumeration of parallel samples.

**Evaluation of the Rapid Screening Assay Using Echovirus 1 in Water Samples**

A preliminary experiment was conducted with sterile beef extract V seeded with Echovirus 1. A non-seeded aliquot of beef extract was also analyzed by plaque assay and the rapid screening assay and confirmed the absence of viruses. The seeded and non-seeded fractions were concentrated by PEG-precipitation, the RNA was extracted followed by serial dilution, and RT-PCR and the rapid screening assay performed. The results indicated that 0.1 – 1 PFU were detected (Table 3.22; enumeration based on plaque assay half of the seeded fraction).

During the following three experiments, six to seven 3-mL aliquots of water concentrates in beef extract V were seeded with different concentrations of Echovirus 1, and one 3-mL aliquot was not seeded (experiment negative control). Serial dilutions of the RNA were performed prior to reverse transcriptase and PCR, leading to multiple Echovirus concentrations. Approach ACM/SL-PRB with colorimetric detection was used on replicate samples (duplicate RNA samples, and duplicate cDNA samples for each reverse transcriptase reaction) and the assay was repeated using the same PCR samples (duplicate microplate assays).

**Table 3.22**
**Detection of Echovirus 1 seeded in sterile beef extract V buffer**

| Number of Echovirus 1 in PCR (PFU) | Signal Intensity | |
|---|---|---|
| | Replicate No.1 | Replicate No.2 |
| 100 | +++ | +++ |
| 10 | ++ | ++ |
| 1 | + | + |
| 0.1 | +/- | - |
| 0.01 | - | - |
| Experiment negative control | - | - |
| Reverse transcriptase negative control | - | - |
| PCR negative control | - | - |

Note: -, no signal; +/-, weak signal; + to ++++, pale to darkest signal.

Results obtained during two of the evaluation experiments are presented in Figure 3.17. The standard deviation was calculated among the four replicate samples processed (two reverse transcriptase reactions and two amplification reactions) and results obtained are presented in Table 3.23. Only the average values of the four replicates were used to prepare the graph. Also, the standard deviations calculated between the replicate samples are low. A predictable progression in signal intensity was observed with increasing Echovirus 1 concentrations and the variation between replicates was low (Table 3.23).

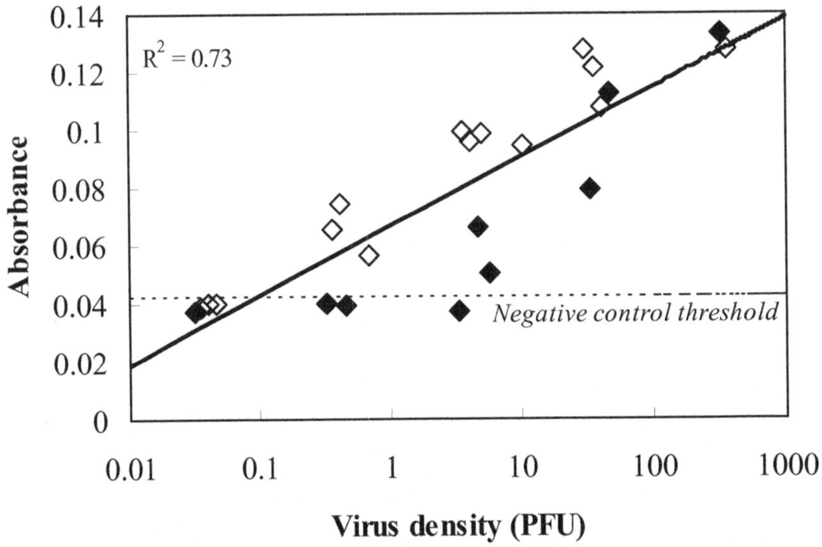


**Figure 3.17. Detection of Echovirus 1 in source water concentrates by the rapid screening assay utilizing the amplicon-coated microplate approach (ACM/SL-PRB) with colorimetric detection. Absorbance was measured at 550 nm and the negative control threshold was the maximum absorbance generated by negative controls. Filled symbols represent the mean of four replicates in one experiment and open symbols are the mean of two replicates from a second experiment.**

**Table 3.23**
**Standard deviation, assay sensitivity and $R^2$ values for replicate evaluation experiments of the enterovirus system**

| Evaluation Experiment | Standard deviation (range) | Sensitivity (PFU) | $R^2$ |
|---|---|---|---|
| Single microplate assay | 0.001 to 0.033 | 2.95 | 0.903 |
| First microplate assay | 0.002 to 0.037 | 0.028 | 0.856 |
| Second microplate assay | 0.001 to 0.042 | 0.073 | 0.846 |
| First microplate assay | 0.001 to 0.037 | 0.004 | 0.831 |
| Second microplate assay | 0.001 to 0.029 | 0.047 | 0.908 |

## EVALUATION OF A COMMERCIAL MICROPLATE HYBRIDIZATION KIT

A commercially available kit uses a method similar to Approach PCM/SL-PRM for the detection of PCR-amplified DNA (Light Diagnostics Oligo Detect; Chemicon International, Inc.) In this kit, the probe is covalently bound to a 96-well microplate. The amplicons are denatured directly in the microplate using an alkaline solution and the biotin-labeled amplicons hybridize to the attached probe. Following hybridization, the microplate is washed, and the colorimetric detection is performed with streptavidin-labeled horse radish peroxidase (HRP) forming a blue colored precipitate at the site of hybridization. The reaction is stopped by the addition of a hydrochloric acid solution, changing the color to yellow, and the color is quantitated using a microplate reader. The kit's instructions provide guidelines to evaluate the signal obtained (positive, negative, or equivocal), based on the average and standard deviation of absorbance values measured at 450 nm. The reported sensitivity is 5 copies of amplified DNA, or 1.5 femtomoles (Chemicon International, Inc.). Since the probe is bound directly to the microplate wall instead of being attached by via a streptavidin-biotin bond, probe pre-coating is not required; the covalent bond may allow a significantly higher concentration of probe to bind to the microplate walls, which may improve the assay sensitivity.

Through a collaborative agreement between Metropolitan and Chemicon, a customized version of the OligoDetect kit was developed for specific detection of *C. parvum*. This customized kit used the same primers and probe sequences that were used for the development and optimization of the rapid screening assay described in this report. The OligoDetect kit is a hybridization capture assay that detects amplified nucleic acid specific to the target microorganism of interest. OligoDetect kits are also available for a variety of pathogens of concern to the water industry, including caliciviruses, enteroviruses, and *Legionella pneumophila*. Similar to the rapid screening assay described in this report, it does not address sampling collection procedures, microorganism concentration methods, or DNA extraction techniques. The kit was tested for sensitivity and specificity and results obtained compared with the rapid screening assay presented in this report. Although a detailed procedure is provided with the kit, it may require optimization to improve signal sensitivity and/or specificity. Therefore several experiments were conducted during which hybridization parameters were modified and signal intensities compared.

### Assay Sensitivity

Experiments were conducted to improve the signal obtained by modifying the incubation conditions of the microplate hybridization and colorimetric detection steps. Signal intensity was read using a microplate reader at dual-wavelength absorbance of 450 nm/650 nm. In the microplate, each sample was processed in triplicate. A summary of the results obtained is presented on Figure 3.18. The results demonstrated that the optimum conditions were hybridization at 40°C for 60 min, with a 45-min incubation at 37°C for the colorimetric detection step (streptavidin-HRP substrate), which provided an assay sensitivity of approximately 1 oocyst (based on serial dilutions of hemocytometer count). Signal intensity increased with increasing hybridization duration and temperature. This trend was not observed with the negative controls, which demonstrates the robustness of the assay.

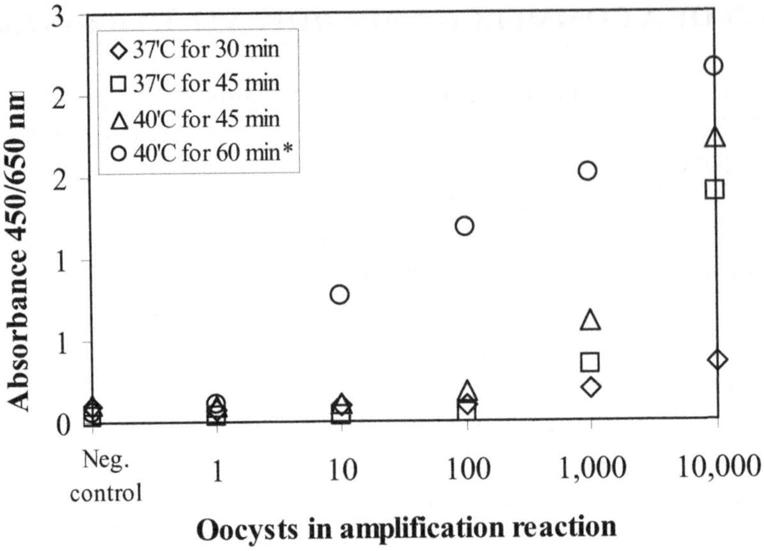


**Figure 3.18. Effect of different hybridization conditions on *C. parvum* detection using the customized OligoDetect kit. For the samples that were hybridized at 40°C for 60 min (*) the colorimetric detection conditions were 45 min at 37°C, whereas all other samples were detected after 30 min at 37°C.**

Three out of the four experiments presented in Figure 3.18 included a colorimetric detection step (streptavidin-HRP) of 30 min at 37°C. The incubation was increased to 45 min at the same temperature (37°C) during the fourth experiment. For the first three experiments, it is possible to attribute the signal improvement to either an increase in hybridization temperature or duration, as the colorimetric detection conditions were similar (37°C for 30 min). However, it is more difficult to determine if the stronger signal obtained during the fourth experiment was attributed to the longer hybridization duration, or colorimetric detection. According to Chemicon's staff, greater variability in signal intensity comes from modifications of the hybridization conditions, and not from the colorimetric detection duration. In addition, the purpose of these experiments was not to evaluate the system responses to changes in assay conditions, but rather to compare the OligoDetect commercial kit to the rapid screening assay presented in this report.

As a means to improve sensitivity, two different reaction volumes were compared during PCR amplification (50 and 100 μL), with the same relative amount of DNA (20% v/v per reaction). Results indicated significantly stronger signal with a reaction volume of 100 μL than with 50 μL (1.31 ± 0.30 fold higher) particularly at the higher *C. parvum* oocysts concentrations tested. This result is important as the rapid screening assay was developed for PCR volumes of 50 μL to decrease costs. However, this indicates that the assay sensitivity can be improved by conducting the amplification in larger volumes.

### Assay Specificity

The assay specificity was examined by testing a selection of microorganisms that are likely to be encountered in source and treated waters, or to interfere with the assay: *E. coli* O157:H7, *S. typhimurium*, *Y. enterocolitica*, *C. perfringens*, *E. cloacae*, *A. hydrophila*, *C. freundii*, *C. muris*, *C. baileyi*, and *S. cerevisiae*. The hybridization conditions were 1 hour at 40°C, with a colorimetric detection (streptavidin-HRP) performed at 37°C for 45 min. Each microorganism was tested in duplicate, and the signal intensity obtained is presented in Figure 3.19. Results demonstrate the specificity of the primers, probe, and assay.

### Assay Duration

Following optimization of the hybridization and colorimetric detection conditions, processing time with the OligoDetect kit was faster (6.3 h from microplate loading with biotin-labeled amplicons to addition of the stop reagent, with 3.8 attended hours), than the preferred Approach ACM/SL-PRB (10.8 h from microplate pre-washing step to signal detection, with 4.3 attended hours). A detail of the assay durations considering analysis of full 96-well microplates, is presented in Table 3.24. These analysis times do not consider reagent preparation nor PCR amplification, which require similar involvement for both assays. The 10.8 h duration for Approach ACM/SL-PRB is a very conservative value as explained in the above paragraphs.

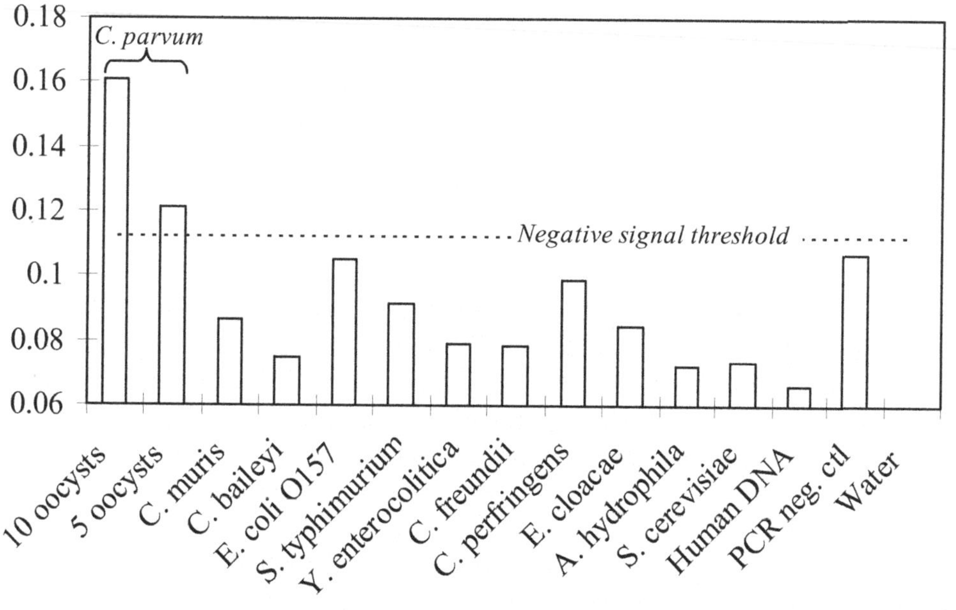


**Figure 3.19. Specificity of the OligoDetect kit customized for *C. parvum* detection. The negative signal threshold was the maximum absorbance value of all negative controls.**

**Table 3.24**
**Comparison of analysis times for OligoDetect kit and rapid screening assay**

| Chemicon International's OligoDetect kit | | Rapid Screening Assay – Approach ACM/SL-PRB | |
|---|---|---|---|
| Operations | Duration (min) | Operations | Duration (min) |
| | | Pre-wash microplate | 60 |
| Load biotin-amplicons into microplate | 60 | Add pre-coating buffer | 15 |
| Add denaturation solution | 15 | Load biotin-amplicons into microplate | 60 |
| Incubation at room temperature | 10 | Incubate at 4°C | 60 |
| | | Microplate washes (2) | 50 |
| Add hybridization buffer | 15 | Add hybridization buffer | 15 |
| Incubation at 40°C | 60 | Incubate at hybridization temperature | 120 |
| Microplate washes (3) | 60 | Microplate washes (2) | 50 |
| Add streptavidin-HRP | 15 | Add antibody (anti-DIG-AP) | 15 |
| Incubation at 37°C | 45 | Incubate at room temperature | 30 |
| Microplate washes (3) | 60 | Microplate washes | 40 |
| Add colorimetric substrate (TMB/E) | 15 | Add colorimetric substrate (NBT/BCIP) | 15 |
| Incubation at 37°C | 10 | Incubate at room temperature | 120 |
| Add stop reagent | 15 | | |
| Total duration | 385 | Total duration | 650 |
| Attended | 230 | Attended | 260 |
| Non-attended | 155 | Non-attended | 390 |

**Controls**

During each experiment, several controls were performed during both the amplification and microplate hybridization steps. During PCR amplification, the negative control consisted of amplification using sterile molecular biology grade water. This negative control was carried throughout the hybridization as well. Several additional negative controls were processed during the hybridization (Figure 3.20), and consisted of amplicons from one of the highest template concentrations tested along with all hybridization reagents, except one, which was substituted by sterile molecular biology grade water: 1) denaturation buffer (w/o denat. buffer), 2) hybridization buffer (w/o hybrid. buffer), 3) streptavidin-HRP conjugate (w/o strept.-HRP), 4) colorimetric substrate (w/o TMB/E), and 5) reaction stop solution (w/o stop reagent). Additional negative controls consisted of wells that did not receive any amplicons (sterile water was substituted), but received all other reagents (water), and "blank" wells which did not receive any amplicon nor reagent, but only sterile water. Most negative controls resulted in signals that were below the negative-signal cut-off values, except for the "w/o denaturation buffer" and "w/o hybridization buffer" (Figure 3.20). These results suggest that some amplicon denaturation occurred even in absence of denaturing reagent (alkaline solution), which is possible considering that this reagent was substituted for sterile molecular biology grade water. Low-salt concentration solutions such as the high quality water used for these controls may lead to unexpected amplicon denaturation. Similarly, hybridization may occur in absence of hybridization buffer as these particular negative controls included the probe already bound to the well walls, and the amplicons during the first step of the assay. The only missing reagent was the hybridization buffer, which was substituted for sterile water. Therefore, some unexpected hybridization may occurred between the amplicons and probe present.

The lyophilized assay positive control (Hybridization Positive Control, HPC) provided with the OligoDetect kit, led to a strong positive signal during each experiment. Under optimized conditions of microplate hybridization and colorimetric detection this signal varied from 2.021 ± 0.098 to 5.357 ± 2.073.

Each sample and control was processed in triplicate in the microplate, except for the assay specificity testing, which was conducted in duplicate. The readings were very close to each other, with standard deviations representing 12.6±13.3% of the average, under optimized conditions of microplate hybridization and colorimetric detection.

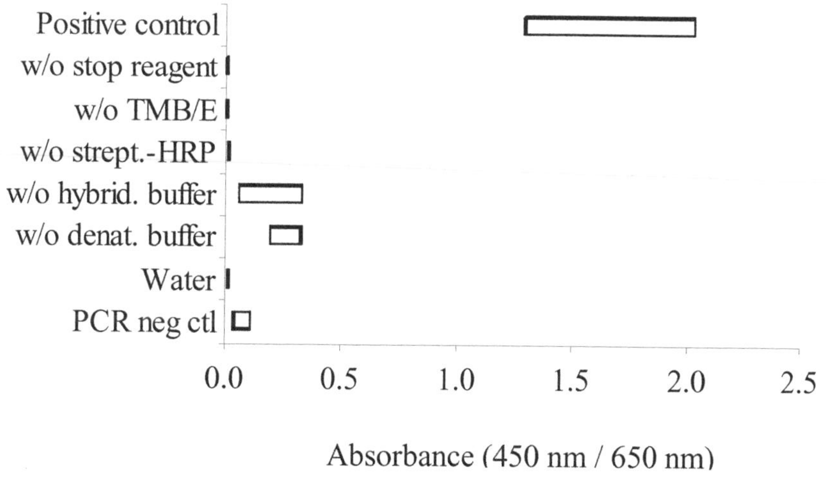


**Figure 3.20. Signal intensity of various controls obtained with the customized OligoDetect kit. The bars represent the range of values obtained from four experiments with hybridization at 37°C for 30 and 45 min and 40°C for 45 and 60 min.**

## CONSIDERATIONS ABOUT DNA EXTRACTION TECHNIQUES

DNA extraction techniques can have a dramatic effect on the quality and quantity of DNA obtained from a sample. Three different DNA extraction techniques were used over the course of this study. Although each method provided a consistent and reproducible extraction of nucleic acid, they may lead to different DNA concentrations, and potentially different assay sensitivities. Because the focus of this study was to develop a rapid screening assay for the detection of pathogens, it was beyond the scope of the project to determine the efficiency of each DNA extraction method, or compare the effectiveness of the rapid screening assay after different DNA extraction procedures. Therefore, all samples used during an experiment were generated during the same PCR, which was conducted using the same DNA extracts. As a result, the different samples processed during the same experiment can be reliably compared. In addition, the overall experimental design eliminated the need to compare data from one experiment to another because each experiment involved modification of a limited number of conditions with comparison to previously tested conditions. The determination of the rapid screening assay sensitivity, specificity and evaluation experiments were all conducted with DNA extracted by the detailed method described in Chapter 2. However, now that the optimum conditions of the rapid screening assay have been defined, they could be used to evaluate the effect of different DNA/RNA extraction techniques on the sensitivity and robustness of the rapid screening assay.

# CHAPTER 4
# CONCLUSIONS AND RECOMMENDATIONS

This project addressed the drinking water industry's need for simple and rapid methods for detection and identification of pathogens in water. The project focused on the development of a detection method and relied on conventional or established sample collection procedures, microorganism recovery and concentration methods, and nucleic acid extraction techniques. An adaptable rapid screening assay was developed for the detection of pathogens in water. The assay was based on an industry-standard 96-well microplate platform that allowed qualitative results to be recorded by simple visual observation and relative quantification by an automated microplate reader. The rapid screening assay consisted of PCR amplification of nucleic acids from target organisms, confirmation by probe hybridization in microplates, and detection of probe-amplicon hybrids using non-radioactive methods. The assay utilized a flexible design that allowed the use of various probe formats and mechanisms of probe attachment, and the assay endpoint was adaptable to both colorimetric and fluorescence-based detection approaches. The assay relies on molecular methods for their high sensitivity and specificity (minimal or no risk of false-positive identification and species specific) and a microplate format for easy and assisted detection, high throughput, and the potential for automation. Proof of principle and demonstration of feasibility was provided by a series of experiments with four selected pathogens (*C. parvum*, *G. duodenalis*, echovirus 1, and *E. coli* O157:H7) but the assay is applicable to any pathogens for which nucleic acid sequence information is available, allowing design of specific primers and probes. The microplate format allows a high number of samples to be processed simultaneously. Therefore, processing time and assay costs are less than those for conventional pathogen detection methods. However, the rapid screening assay is not truly quantitative since it provides either presence/absence results or relative quantification. Also it does not provide information on the viability and/or infectivity of the microorganisms when used simply as a detection tool for pathogens recovered directly from water. Nevertheless, the assay developed by this project offers many advantages as a pathogen-screening tool. Specific conclusions of this project include the following:

1. The rapid screening assay is a useful screening tool that may be used to eliminate negative samples, which constitute the majority of drinking water samples. More extensive analyses such as microbial enumeration and viability infectivity assays can then be focused solely on the positive samples.
2. The microplate format allows many samples to be processed simultaneously without increasing analysis time. Consequently, processing time and costs are substantially reduced compared to conventional culture and microscopy-based microbial detection analyses.
3. Two assay formats were evaluated involving either probe-coated microplates (PCM) or amplicon-coated microplates (ACM). In molecular based assays, the amplified target, hybridization conditions, and the detection signal or moiety are the three key components that must be optimized. Therefore, different combinations of amplified target, microplate hybridization and end-point detection approaches were systematically examined with regards to sensitivity, specificity and robustness. Both assay formats provided similar results in terms of assay sensitivity and duration. However, the ACM format was selected as the most appropriate because it was less expensive, more readily adapted to a wide

range of organisms, less sensitive to variable assay conditions, and had greater flexibility in analysis options compared to the PCM approach.

4. As the first component of the assay, label incorporation into amplicons was achieved without modifying reported amplification conditions. A single round of amplification (35 – 40 cycles) provided sufficient sensitivity and amplicon purification was not necessary.
5. The ACM approach was only slightly affected by changes in the different assay parameters. The critical steps were amplicon denaturation and pre-coating techniques, sample volume, sodium concentration during hybridization, and sufficient contact with the colorimetric substrate. Considerable probe optimization was necessary for the PCM approach suggesting that it is less practical for transferring to other organisms.
6. For *C. parvum*, a published primer/probe combination was used for the assay and a detection sensitivity of less than 10 oocysts was achieved (based on serial dilutions of hemocytometer-enumerated oocyst stock preparations).
7. With echovirus 1, the sensitivity of ACM approach was less than 0.02 plaque forming units (PFU; equivalent to less than 20 virus particles) using a primer/probe combination specific for all human enteroviruses including poliovirus, coxsackie A and B viruses, and echoviruses.
8. The assay was easily adapted for the detection of other microorganisms such as *G. duodenalis* and *E. coli* O157:H7. Published primer/probe combinations and amplification conditions were used without further modification. The sensitivity was 9 cysts and 10 colony-forming units (CFU) for *G. duodenalis* and *E. coli* O157:H7, respectively.
9. Evaluations of assay sensitivity with seeded environmental water samples demonstrated sensitivities of 2 – 9 *C. parvum* oocysts (based on immunofluorescent enumeration) and 0.004 – 0.073 PFU of echovirus 1 (equivalent to 4 – 73 virions), with good reproducibility between replicates.
10. Evaluation of a commercially available hybridization-capture kit (OligoDetect) customized for *C. parvum* and incorporating the same primer/probe combination used for the rapid screening assay demonstrated sensitive and specific detection. The protocol for the commercial kit was more streamlined and so required less analysis time (6.4 h) compared to the assay developed for the project (10.8 h), although considerable optimization is still required for successful application of the kit. However, commercialization of a similar technology demonstrates the value and utility of the type of assay developed for this project.

## RECOMMENDATIONS AND FUTURE RESEARCH

The demonstration of feasibility of a microplate-based rapid screening assay provided by this project indicates that the water industry should continue to pursue and fund exploratory research into the development of innovative pathogen detection methods. Such techniques may be used for routine screening of drinking water, during outbreak situations, or to rule out (or in) intentional contamination. All of these situations require pathogen detection tools that can generate results, even if only presumptive results, in as short a turn around time as possible. Although considerable assay optimization and evaluation were performed during the course of this project, several issues that warrant further evaluation were identified:

1. The microplate-based assay was developed as a presence/absence screening tool with a colorimetric end-point for the detection of pathogens in water. Assay results were obtained by simple visual observation and a qualitative scoring scale ranging from no signal to dark signal. Although this scale is practical and easy to implement, interpretation may be subjective with pale and weak positive signals. Therefore, relative quantification was achieved through the use of an automated plate reader at a wavelength of 550 nm. However, an assessment of possible interferences by compounds other than the specific colorimetric precipitate is necessary to ensure accurate plate reader results. In addition, more extensive evaluations with precisely enumerated seed organisms need to be performed to improve the quantitative aspects of the assay and to more fully evaluate the sensitivity. In the case of *C. parvum*, oocysts enumerated by flow cytometry would be appropriate as seed organisms.
2. The assay was developed as a detection tool to use in combination with PCR (or other nucleic acid amplification methods) following recovery of organisms from water and extraction of nucleic acids. However, the utility of the assay would be increased by coupling it to direct nucleic extraction techniques so that potentially all pathogens can be screened from a single nucleic acid extraction thereby uncoupling the assay from time-consuming and often inefficient conventional microbial recovery methods.
3. Assay standardization, validation and round-robin testing are required to assess the transferability of the technology. In addition, the assay would benefit from streamlining and simplification in some of the processing steps, resulting in greater transferability. Specifically, further experiments involving shorter incubation periods are necessary to determine the minimum possible assay duration.
4. Seeded studies using a variety of source water types should be conducted to assess the robustness of the assay and to examine possible interferences.
5. Since the microplate assay is a detection tool, it can be used for applications other than pathogen detection. As an example, it could be used as an alternative to gel electrophoresis and membrane hybridization in cell culture-based viral and protozoal infectivity assays that use PCR or RT-PCR to detect infections. This and other potential applications of the assay should be explored in more detail.

# APPENDIX A
# STANDARD OPERATING PROCEDURE FOR MICROPLATE HYBRIDIZATION - APPROACH PCM/ML-DUTP

**STANDARD OPERATING PROCEDURE FOR MICROPLATE HYBRIDIZATION - APPROACH PCM/ML-DUTP**

40 μL sample (PCR amplicons), 96-well microplate, 300 μL per well maximum.

Pre-wash: Prepare a microplate pre-washing solution as follows:

In each well add:
- 15 μL 1.0 M Tris-HCl (pH 7.5) (50 mM final concentration)
- 6 μL 1% SDS (0.02% final concentration)
- 279 μL sterile Milli-Q water

Add 300 μL of pre-washing solution per well, and incubate 5 min at room temperature on Belly Dancer (agitation). Flick into sink, and remove excess liquid using a clean absorbent paper.

Rinse: Add 300 μL sterile Milli-Q water per well, incubate 5 min at room temperature on Belly Dancer. Flick into sink, and remove excess liquid using a clean absorbent paper.
Repeat rinsing 1 more time (2 rinses total).

Pre-coating: Prepare pre-coating solution as follows:

In each well add:
- 75 μL 20x SSC (6x SSC final concentration)
- 25 μL 10% blocking solution (1% final concentration)
- 5 μL 1% SDS (0.02% final concentration)
- 2 μL 5.0 nM biotin-labeled probe (40 pM final concentration)
- 143 μL sterile Milli-Q water

Add 250 μL of pre-coating solution per well, and incubate 30 min room temperature on Belly Dancer. Flick into sink, and remove excess liquid using a clean absorbent paper.

Wash: Prepare a washing solution of 5x SSC, and 20 mM Tris-HCl (pH 7.5).
Fill each well with washing solution. Let sit 5 min at room temperature.
Flick into sink, and remove excess liquid using a clean absorbent paper.
Repeat washing 1 more time (2 washes total).

Heat denaturation: Heat samples at 95°C for 5 min, then at 99°C for 1 min.
Immediately, put samples on crushed ice/water (~0°C) (keep from re-annealing) for 5 min.
Keep cold until ready to use.

Hybridization: Prepare hybridization solution as follows:

In each well add:
- 75 μL 20x SSC (6x SSC final concentration)
- 25 μL 10% blocking solution (1% final concentration)
- 5 μL 1% SDS (0.02% final concentration)
- 105 μL sterile Milli-Q water

Cool solution to 4°C. Add 210 μL of cold hybridization solution per well, add 40 μL of sample (work at 4°C), and incubate at hybridization temperature, 2 h. Agitate gently during incubation. Flick into sink, and remove excess liquid using a clean absorbent paper.

Wash: Prepare a washing solution of 5x SSC, and 20 mM Tris-HCl (pH 7.5). Warm up to hybridization temperature. Fill each well with warm washing solution (290 μL). Let sit 10 min at washing temperature. Flick into sink, and remove excess liquid using a clean absorbent paper. Repeat washing 1 more time (2 washes total).

Antibody: Prepare a conjugate solution of 1% (w/v) BSA in Tris buffer (50 μL of 20 mg/mL BSA in 100 mL Tris buffer) containing a 1:1000 dilution of alkaline phosphatase-labeled anti-DIG antibody (for a 96-well microplate, use 30 μL antibody in 30 mL Tris-BSA buffer). Fill each well (300 μL), and incubate 30 min at room temperature on Belly Dancer. Flick into sink, and remove excess liquid using a clean absorbent paper.

Wash: Fill each well with 1% (w/v) BSA in Tris buffer (300 μL). Incubate 5 min at room temperature on Belly Dancer. Flick into sink, and remove excess liquid using a clean absorbent paper. Repeat washing 1 more time (2 washes total).

Detection: Prepare a substrate solution of NBT/BCIP by adding 600 μL of NBT/BCIP stock solution to 30 mL Tris buffer (for a 96-well microplate). Wrap substrate solution in aluminum foil to preserve from light. Fill each well (300 μL). Within 15 min of incubation in darkness, a dark-purple color will confirm the presence of the tested organism. The color precipitate start to form within a few min, and the reaction is usually complete after 1 d. Do not shake or mix while color is developing. Record results after 1h00 and 2h00 of incubation.
Flick into waste container, and remove excess liquid using a clean absorbent paper.

Wash: Fill each well with 1X TE buffer (300 μL). Incubate 5 min at room temperature in darkness. Flick into waste container, and remove excess liquid using a clean absorbent paper. Repeat washing 1 more time (2 washes total). Verify readings. Dry well.
Wrap microplate in aluminum foil, and store at 4°C.
Note: Color fades upon drying, but can be revitalized by adding TE buffer into wells.

Picture: Place microplate on light table. Use yellow filter #8, f > 32, and speed 125.

**Preparation of solutions**

1% SDS (lauryl sulfate): Prepare a 1/10 dilution of 10% SDS solution (Sigma, #L-4522) in sterile Milli-Q water.

1X TE buffer: Prepare a 1/100 dilution of 100X Tris-EDTA solution (Sigma #T-9285) in sterile Milli-Q water.

First washing solution:
125 mL 20xSSC (final concentration of 5xSSC)
1.211 g Tris (final concentration of 20 mM)
Adjust to pH 7.50 with 1 M HCl
Complete to 500 mL with sterile Milli-Q water
Sterilize by autoclaving.

Tris buffer: 100 mM Tris (6.055 g)

50 mM $MgCl_2$ (25 mL of 1 M solution)
100 mM NaCl (2.922 g)
Adjust to pH 9.50 with HCl 1 M
Complete to 500 mL with sterile Milli-Q water
Filter sterilize through 0.22 μm (autoclaving forms a flaky precipitate!).

Alkaline phosphatase-labeled anti-DIG antibody: Roche Applied Science.

NBT/BCIP substrate solution: Roche Applied Science.

Blocking solution: Roche Applied Science, 10% in sterile deionized water.

# APPENDIX B
# STANDARD OPERATING PROCEDURE FOR MICROPLATE HYBRIDIZATION - APPROACH ACM/SL-PRB

**STANDARD OPERATING PROCEDURE FOR MICROPLATE HYBRIDIZATION - APPROACH ACM/SL-PRB**

40 μL sample (PCR amplicons), 96-well microplate, 300 μL per well maximum.

Pre-wash: Prepare a microplate pre-washing solution as follows:
In each well add: 15 μL 1.0 M Tris-HCl (pH 7.5) (50 mM final concentration)
6 μL 1% SDS (0.02% final concentration)
279 μL sterile Milli-Q water
Add 300 μL of pre-washing solution per well, and incubate 5 min at room temperature on Belly Dancer (agitation). Flick into sink, and remove excess liquid using a clean absorbent paper.

Rinse: Add 300 μL sterile Milli-Q water per well, incubate 5 min at room temperature on Belly Dancer. Flick into sink, and remove excess liquid using a clean absorbent paper.
Repeat rinsing 1 more time (2 rinses total).

Heat denaturation: Heat samples at 95°C for 5 min, then at 99°C for 1 min.
Immediately, put samples on crushed ice/water (~0°C) (keep from re-annealing) for 5 min.
Keep cold until ready to use.

Pre-coating: Prepare pre-coating solution as follows:
In each well add: 62.5 μL 20x SSC (5x SSC final concentration)
25 μL 10% blocking solution (1% final concentration)
5 μL 1% SDS (0.02% final concentration)
117.5 μL sterile Milli-Q water
40 μL sample
Cool solution to 4°C. Add 210 μL of cold pre-coating solution per well, add 40 μL of sample (work at 4°C), and incubate at 4°C for 1 h.
Flick into sink, and remove excess liquid using a clean absorbent paper.

Wash: Prepare a washing solution of 5x SSC, and 20 mM Tris-HCl (pH 7.5). Cool to 4°C.
Fill each well with 300 μL cold washing solution (work at 4°C). Let sit 10 min at 4°C.
Flick into sink, and remove excess liquid using a clean absorbent paper.
Repeat washing 1 more time (2 washes total).

Hybridization: Prepare hybridization solution as follows:
In each well add: 62.5 μL 20x SSC (5x SSC final concentration)
25 μL 10% blocking solution (1% final concentration)
5 μL 1% SDS (0.02% final concentration)
2 μL 5.0 nM DIG-labeled probe (40 pM final concentration)
155.5 μL sterile Super-Q water
Add 250 μL of hybridization solution per well, and incubate in darkness at hybridization temperature for 2 h. Agitate gently during incubation.
Flick into sink, and remove excess liquid using a clean absorbent paper.

Wash: Prepare a washing solution of 5x SSC, and 20 mM Tris-HCl (pH 7.5). Warm up to hybridization temperature. Fill each well with warm washing solution (290 μL).
Let sit 10 min at washing temperature, in darkness.
Flick into sink, and remove excess liquid using a clean absorbent paper.
Repeat washing 1 more time (2 washes total).

Antibody: Prepare a conjugate solution of 1% (w/v) BSA in Tris buffer (50 μL of 20 mg/mL BSA in 100 mL Tris buffer) containing a 1:1000 dilution of alkaline phosphatase-labeled anti-FITC antibody (for a 96-well microplate, use 30 μL antibody in 30 mL Tris-BSA buffer).
Fill each well (300 μL), and incubate 30 min in darkness at room temperature on Belly Dancer.
Flick into sink, and remove excess liquid using a clean absorbent paper.

Wash: Fill each well with 1% (w/v) BSA in Tris buffer (300 μL). Incubate 5 min at room temperature in darkness on Belly Dancer. Flick into sink, and remove excess liquid using a clean absorbent paper. Repeat washing 1 more time (2 washes total).

Detection: Prepare a substrate solution of NBT/BCIP by adding 600 μL of NBT/BCIP stock solution to 30 mL Tris buffer (for a 96-well microplate). Wrap substrate solution in aluminum foil to preserve from light. Fill each well (300 μL). Within 15 min of incubation in darkness, a dark-purple color will confirm the presence of the tested organism. The color precipitate start to form within a few min, and the reaction is usually complete after 1 d. Do not shake or mix while color is developing. Record results after 1h00 to 2h00 of incubation.
Flick into waste container, and remove excess liquid using a clean absorbent paper.

Wash: Fill each well with TE buffer (300μL). Incubate 5 min at room temperature. Flick into waste container, and remove excess liquid using a clean absorbent paper.
Repeat washing 1 more time (2 washes total). Verify readings. Dry well.
Wrap microplate in aluminum foil, and store at 4°C.
Note: Color fades upon drying, but can be revitalized by adding TE buffer into wells.

Picture: Place microplate on light table. Use yellow filter #8, f > 32, and speed 125.

**Preparation of solutions**

1% SDS (lauryl sulfate): Prepare a 1/10 dilution of 10% SDS solution (Sigma, #L-4522) in sterile Milli-Q water.

1X TE buffer: Prepare a 1/100 dilution of 100X Tris-EDTA solution (Sigma #T-9285) in sterile Milli-Q water.

First washing solution:
125 mL 20xSSC (final concentration of 5xSSC)
1.211 g Tris (final concentration of 20 mM)
Adjust to pH 7.50 with 1 M HCl
Complete to 500 mL with sterile Milli-Q water, Sterilize

Tris buffer: 100 mM Tris (6.055 g)

50 mM $MgCl_2$ (25 ml of 1 M solution)
100 mM NaCl (2.922 g)
Adjust to pH 9.50 with HCl 1 M
Complete to 500 mL with sterile Milli-Q water
Filter sterilize through 0.22 μm (autoclaving forms a flaky precipitate!).

Alkaline phosphatase-labeled anti-DIG antibody: Roche Applied Science.

NBT/BCIP substrate solution: Roche Applied Science.

Blocking solution: Roche Applied Science, 10% in sterile deionized water.

## REFERENCES


Abbaszadegan, M., M.S. Huber, C.P. Gerba, and I.L. Pepper. 1997. Detection of Viable *Giardia* Cysts by Amplification of Heat Shock-Induced mRNA. *Appl. Environm. Microbiol.*, 63:324-328.

APHA. 1998. Standard Methods for the Examination of Water and Wastewater, 20th edition. APHA, AWWA and WEF, Washington, D.C.

Barwick, R.S., D.A. Levy, G.F. Craun, M.J. Beach, and R.L. Calderon. 2000. Surveillance for Waterborne Disease Outbreaks - United States, 1997-1998, In: CDC Surveillance Summaries. *Morbid. Mortal. Weekly Rep.*, May 26, 49:SS-4:1-35.

Blewett, D.A.,S.E. Wright, D.P. Casemore, N.E. Booth, and C.E. Jones. 1993. Infective Dose Size Studies on *Cryptosporidium parvum* Using Gnotobiotic Lambs. *IAWQ*, 27:3-4:61-64.

Bon, F., H. Giraudon, C. Sancey, C. Barringer, M. Joannes, P. Pothier, and E. Kohli. 2004. Development and Evaluation of a New Commercial Test Allowing the Simultaneous Detection of Noroviruses and Sapoviruses by Reverse Transcription-PCR and Microplate Hybridization. *Jour. Clin. Microbiol.*, 42:2218-2220.

Brown, T.A. 1991. Molecular Biology - LABFAX. BIOS Scientific Publishers Ltd, Oxford, UK.

Craun, G.F. 1988. Surface Water Supplies and Health. *Jour. AWWA*, 80:40-52.

Craun, G.F. 1990. Methods for Investigation and Prevention of Waterborne Disease Outbreaks, EPA/600/1-90/005a, USEPA Office of Research and Development.

Craun, G.F. 1992. Waterborne Disease Outbreaks in the United States of America: Causes and Prevention. *World Health Stat. Q.*, 45(2-3):192-199.

Craun, G.F., R.L. Calderon, and N. Nwachuku. 2003. Causes of Waterborne Outbreaks Reported in the United States, 1991-1998. In *Drinking Water and Infectious Disease: Establishing the Links*, Edited by P.R. Hunter, M. Waite, and E. Ronchi. Boca Raton, Florida: CRC Press.

Cromeans, T.L., J. Narayanan, K. Jung, G. Ko, D. Wait, and M.D. Sobsey. 2004. *Development of Molecular Methods to Detect Infectious Viruses in Water*. Denver, Colo.: Awwa Research Foundation.

De Leon, R., C. Shieh, R.S. Baric, and M.D. Sobsey. 1990. Detection of Enteroviruses and Hepatitis A Virus in Environmental Samples by Gene Probes and Polymerase Chain Reaction. *AWWA - Water Quality Technology Conference*, San Diego, CA.

De Leon, R., and M.D. Sobsey. 1991. Methods for Virus Detection in Water. In: *Monitoring Water in the 1990's: Meeting New Challenges, ASTM STP 1102*. Ed. J.R. Hall and G.D. Glysson, American Society for Testing and Materials, Philadelphia, PA, 400-421.

Erhardt, A., S. Schaefer, N. Athanassiou, M. Kann, and W.H. Gerlich. 1996. Quantitative Assay of PCR-Amplified Hepatitis B Virus DNA Using a Peroxidase-Labelled DNA Probe and Enhanced Chemiluminescence. *Jour. Clin. Microbiol.*, 34:1885-1891.

Egholm, M., O. Buchardt, L. Christensen, C. Behrens, C. Freier, D.A. Driver, R.H. Berg, S.K. Kim, B. Norden, and P.E. Nielson. 1993. PNA Hybridizes to Complementary Oligonucleotides Obeying Watson-Crick Hydrogen Bonding Rules. *Nature*, 365:566-568.

Fayer, R., U. Morgan, and S.J.Upton. 2000. Epidemiology of *Cryptosporidium*: Transmission, Detection, and Identification. *Int. Jour. Parasitol.*, 30:1305-1322.

Feng, P. 1995. *Escherichia coli* Serotype 0157:H7: Novel Vehicles of Infection and Emergence of Phentypic Variants. *Emerging Infec. Dis.,* 2:47.

Gerhardt, P., R.G.E. Murray, W.A. Wood, and N.R. Krieg. 1994. Methods for General and Molecular Bacteriology. ASM publishers, Washington D.C.

Hames, B.D. 1986. Nucleic Acid Hybridization – A Practical Approach. IRL Press.

Howard, G.C. 1993. *Methods in Nonradioactive Detection.* Appleton & Lange, East Norwalk, CT.

Hrudey, S.E. and E.J. Hrudey. 2004. *Safe Drinking Water: Lessons from recent Outbreaks in Affluent Nations*. London: IWA Publishing.

Hunter, P.R., and Q Syed. 2001. Community Surveys of Self-reported Diarrhoea can Dramatically Overestimate the Size of Outbreaks of Waterborne Cryptosporidiosis. *Wat. Sci. Technol.*, 43:27-30.

Jiang, X., P.-W. Huang, V.R. Land, and D.O. Matson. 2001. *Detection of Caliciviruses in Water Samples*. Denver, Colo.: Awwa Research Foundation.

Kramer, M.H., B.L. Herwaldt, G.F. Craun, R.L. Calderon, and D.D. Juranek. 1996. Surveillance for Waterborne-Disease Outbreaks—United States, 1993-1994. MMWR, 45:SS-1:1-33.

Lang, A.L., Y.L. Tsai, C.L. Mayer, K.C. Patton, and C.J. Palmer. 1994. Multiplex PCR for Detection of the Heat-Labile Toxin Gene and Shiga-Like Toxin I and II Genes in *Escherichia coli* Isolated from Natural Water. *Appl. Environm. Microbiol.*, 60(9):3145-3149.

Longo, M.C., M.S. Berninger, and J.L. Hartley. 1990. Use of Uracil DNA Glycosylase to Control Carry-over Contamination in Polymerase Chain Reactions. *Gene*, 93:125-128.

MacKenzie, W.R., N.J. Hoxie, M.E. Proctor, M.S. Gradus, K.A. Blair, D.E. Peterson, J.J. Kazmierczak, D.G. Addiss, K.R. Fox, J.B. Rose, and J.P. Davis. 1994. A Massive Outbreak in Milwaukee of *Cryptosporidium* Infection Transmitted through the Public Water Supply. *N. Engl. J. Med.*, 331:161-167.

Melnick, J.L., H.A. Wenner, and C.A. Phillips. 1979. Enteroviruses. In: *Diagnostic Procedures for Viral, Rickettsial and Chlamydial Infections*, 5$^{th}$ edition. Ed. E.H. Lennette and N.J. Schmidt, American Public Health Association, Washington D.C., 471-534.

Monday, S.R., T.S. Whittman, and P.C.H. Feng. 2001. Genetic and Evolutionary Analysis of Mutations in the gusA Gene That Cause the Absence of β-Glucuronidase Activity in *Escherichia coli* O157:H7. *Jour. Infect. Dis.*, 184:918-921.

O'Connor, D.R. 2002. Report of the Walkerton Inquiry – The Events of May 2000 and Related Issues. Queen's Printer for Ontario, Published by Ontario Ministry of the Attorney General.

Okhuysen, P.C., C.L. Chappell, J.H. Crabb, C.R. Sterling, and H.L. DuPont. 1999. Virulence of Three Distinct *Cryptosporidium parvum* Isolates for Healthy Adults. *Jour. Infect. Dis.*, 180:1275-1281.

Olsvik, O., and N.A. Strockbine. 1993. PCR Detection of Heat-Stable, Heat-Labile, and Shiga-Like Toxin Genes in *Escherichia coli.* In: *Diagnostic Molecular Microbiology: Principles and Applications*. American Society for Microbiology, Washington, D.C., 271-276.

Ou, C.Y., S. Kusk, S.W. Mitchell, D.H. Mack, J.J. Snisnsky., J.W. Krebs, P. Feorino, D. Warfield, and G. Schochetman. 1988. DNA Amplification for Direct Detection of HIV-1 in DNA of Peripheral Blood Mononuclear Cells. *Science*, 239:295-297.

Reynolds, K.A., C.P. Gerba, I.L. Pepper, and M. Abbaszadegan. 2001. Rapid PCR-Based Monitoring of Infectious Enteroviruses in Drinking Water. Denver, Colo.: Awwa Research Foundation.

Rochelle, P.A. 2001. Detection of Protozoa in Environmental Water Samples. In *Environmental Molecular Microbiology: Protocols and Applications*, Edited by P.A. Rochelle. Wymondham: Horizon Scientific Press.

Rochelle, P.A., R. De Leon, M.H. Stewart, and R.L. Wolfe. 1997a. Comparison of Primers and Optimization of PCR Conditions for Detection of *Cryptosporidium parvum* and *Giardia lamblia* in Water. *Appl. Environm. Microbiol.*, 63(1):106-114.

Rochelle, P.A., D.M. Ferguson, T.J. Handojo, R. De Leon, M.H. Stewart, and R.L. Wolfe. 1997b. An Assay Combining Cell Culture with Reverse Transcriptase PCR to Detect and Determine the Infectivity of Waterborne *Cryptosporidium parvum. Appl. Environ. Microbiol.*, 63(5):2029-2037.

Rochelle, P.A., M.M. Marshall, J.R. Mead, A.M. Johnson, D.G. Korich, J.S. Rosen, and R. De Leon. 2002. Comparison of In-vitro Cell Culture and a Mouse Assay for Measuring Infectivity of *Cryptosporidium parvum*. *Appl. Environ. Microbiol.*, 68:3809-3817.

Sambrook, J.E., E.F. Fritsch, and T. Maniatis. 1989. *Molecular Cloning: A Laboratory Manual*, 2nd edition. Cold Spring Harbor Laboratory Press, Cold Spring Harbor, N.Y.

Schmidt M., B. Frey, K. Kaluza, and H. Sobek. 1996. Application of Heat-labile Uracil-DNA-Glycosylase in Improved Carryover Prevention Technique. *Biochemica*, 2:13-15.

Schwab, K.J., R. De Leon, and M.D. Sobsey. 1995. Concentration and Purification of Beef Extract Mock Eluates from Water Samples for the Detection of Enteroviruses, Hepatitis A Virus, and Norwalk Virus by Reverse-Transcription-PCR. *Appl. Environ. Microbiol.*, 61(2):531-537.

Schwab, K.J., R. De Leon, and M.D. Sobsey. 1996. Immunoaffinity Concentration and Purification of Waterborne Enteric Viruses for Detection by Reverse Transcriptase PCR. *Appl. Environ. Microbiol.*, 62(6):2086-2094.

Scott Eastman, P., E. Boyer, L. Mole, J. Kolberg, M. Urdea, and M. Holodniy. 1995. Nonisotopic Hybridization Assay for Determination of Relative Amounts of Genotypic Human Immunodeficiency Virus Type 1 Zidovudine Resistance. *Jour. Clin. Microbiol.*, 33:2777-2780.

Shaw, P.K. 1977. A Communitywide Outbreak of Giardiasis with Evidence of Transmission by a Municipal Water Supply. *Ann. Intern. Med.*, 87:4:426-432.

Stinear, T.P., P.D.R. Johnson, and J.K Davies. 2001. Direct Detection of Mycobacteria in the Environment Using Sequence Capture-PCR. In *Environmental Molecular Microbiology: Protocols and Applications*, Edited by P.A. Rochelle. Wymondham: Horizon Scientific Press.

Takeda, N., K. Sakae, M. Agboatwala, S. Isomura, R. Hondo, and S. Inouye. 1994. Differentiation Between Wild and Vaccine-Derived Strains of Poliovirus by Stringent Microplate Hybridization of PCR Products. *Jour. Clin. Microbiol.*, 32:202-204.

Tanaka, M., S. Onoe, T. Matsuba, S. Katayama, M. Yamanaka, H. Yonemichi, K. Hiramatsu, B.K. Baek, C. Sugimoto, and M. Onuma. 1993. Detection of *Theileria sergenti* Infection in Cattle by Polymerase Chain Reaction Amplification of Parasite-Specific DNA. *Jour. Clin. Microbiol.*, 31:2565-2569.

U S. Environmental Protection Agency. 1989. Drinking Water; National Primary Drinking Water Regulations; Filtration and Disinfection; Turbidity, *Giardia lamblia*, Viruses, *Legionella*, and Heterotrophic Bacteria; Final Rule. *Fed. Reg.*, 54:124:27486 (June 29).

U.S. Environmental Protection Agency. 1996. National Primary Drinking Water Regulation: Monitoring Requirements for Public Drinking Water Supplies: *Cryptosporidium*, *Giardia*, Viruses, Disinfection By-products, Water Treatment Plant Data and Other Information Requirements. Final Rule. *Fed. Reg.*, 61:94:24354-24388 (May 14).

U.S. Environmental Protection Agency. 1998. Contaminant Candidate List. *Fed. Reg.*, 63:40:10273-10287 (March 2).

U.S. Environmental Protection Agency. 1999. Method 1623: *Cryptosporidium* and *Giardia* in Water by Filtration/IMS/FA. Office of Water, Washington, D.C. (April).

# ABBREVIATIONS

| | |
|---|---|
| ACM | amplicon coated microplate |
| AP | alkaline phosphatase |
| APHA | American Public Health Association |
| ASTM | American Society for Testing and Materials |
| ATCC | American Type Culture Collection |
| AwwaRF | Awwa Research Foundation |
| | |
| BGMK | buffalo green monkey kidney-derived |
| BIOTEG | biotin-labeled DNA fragment with a 15-carbon chain link |
| BSA | bovine serum albumin |
| | |
| °C | degrees Celsius |
| *C. baileyi* | *Cryptosporidium baileyi* |
| CDHS-VRDL | California Department Health Services Viral and Rickettsial Disease Laboratory |
| CFU | colony forming unit |
| *C. parvum* | *Cryptosporidium parvum* |
| *C. muris* | *Cryptosporidium muris* |
| CPE | cytopathic effects |
| CRW | Colorado River water |
| CTAB | hexadecyltrimethylammonium bromide |
| | |
| DAPI | 4,6-diamidino-2-phenylindole |
| DIG | digoxigenin |
| | |
| EAEC | enteroadherent-aggregative *Escherichia coli* |
| *E. coli* | *Escherichia coli* |
| EDTA | ethylenediaminetetraacetic acid |
| EHEC | enterohemorrhagic *Escherichia coli* |
| EIEC | enteroinvasive *Escherichia coli* |
| ent | enterovirus |
| EPEC | enteropathogenic *Escherichia coli* |
| ETEC | enterotoxigenic *Escherichia coli* |
| | |
| FITC | fluorescein isothiocyanate |
| | |
| *G. duodenalis* | *Giardia duodenalis* |
| *G. muris* | *Giardia muris* |
| | |
| h | hour |
| HCl | hydrochloric acid |
| HRP | horse radish peroxidase |
| Hyb. | hybridization |

| | |
|---|---|
| ICR | Information Collection Rule |
| IFA | indirect immunofluorescent assay |
| IMS | immuno-magnetic separation |
| | |
| KCl | potassium chloride |
| *K. pneumoniae* | *Klebsiella pneumoniae* |
| *K. trevisanii* | *Klebsiella trevisanii* |
| *K. oxytoca* | *Klebsiella oxytoca* |
| | |
| L | liter |
| | |
| M | molar |
| $MgCl_2$ | magnesium chloride |
| min | minute |
| mL | milliliter |
| ML | multiple labels |
| mM | millimolar |
| MOI | multiplicity of infection |
| MUG | 4-methylumbelliferyl-β-*D*-glucuronide |
| μg | microgram |
| μL | microliter |
| μM | micromolar |
| | |
| NaCl | sodium chloride |
| NaOH | sodium hydroxide |
| NBT/BCIP | nitroblue tetrazolium salt / 5-bromo-4-chloro-3-indolyphosphate |
| neg. | negative |
| ng | nanogram |
| nm | nanometer |
| | |
| PAC | Project Advisory Committee |
| PBS | phosphate buffered saline |
| PCM | probe coated microplate |
| PCR | polymerase chain reaction |
| PEG | polyethylene glycol |
| PFU | plaque forming unit |
| pM | picomolar |
| PNA | peptide nucleic acid |
| PRB | probe |
| | |
| RD | rhabdomyosarcoma |
| RT | reverse transcriptase |
| | |
| SD | standard deviation |
| SDS | sodium dodecyl sulfate (lauryl sulfate) |

| | |
|---|---|
| sec | second |
| SL | single label |
| SLT | Shiga-like toxin |
| sp. | species |
| SPW | State Project water |
| SSC | saline sodium citrate |
| | |
| $T_d$ | disassociation temperature |
| $T_m$ | melting temperature |
| TMB/E | 3,3',5,5'-tetramethylbenzidine/Enhancer |
| TSB | tryptic soy broth |
| | |
| U | unit |
| UNG | uracil DNA-glycosylase |
| U.S. | United States |
| USEPA | United States Environmental Protection Agency |
| | |
| v/v | volume/volume |
| | |
| w/v | weight/volume |

IWA Publishing's authorised EU representative for General Product Safety Regulations is Diane D'Arras, 15 rue Duret, 75116 Paris, France, e-mail: safety@iwap.co.uk.

Printed and bound by CPI Group (UK) Ltd, Croydon, CR0 4YY
17/09/2026
02218799-0014

